The Student Edition of
MICRO-CAP III ™

Limited Warranty

Addison-Wesley warrants that the Student Edition of MICRO-CAP III ("the program") will substantially conform to the published specifications and to the documentation during the period of 90 days from the date of original purchase, provided that it is used on the computer hardware and with the operating system for which it was designed. Addison-Wesley also warrants that the magnetic media on which the program is distributed and the documentation are free from defects in materials and workmanship during the period of 90 days from the date of original purchase. Addison-Wesley will replace defective media or documentation or correct substantial program errors at no charge, provided you return the item with dated proof of purchase to Addison-Wesley within 90 days of the date of original purchase. If Addison-Wesley is unable to replace defective media or documentation or correct substantial program errors, your license fee will be refunded. These are your sole remedies for any breach of warranty.

Except as specifically provided above, Addison-Wesley makes no warranty or representation, either express or implied, with respect to this program, documentation or media, including their quality, performance, merchantability, or fitness for a particular purpose. Spectrum makes no warranty or representation, either express or implied, with respect to this program, documentation or media, including their quality, performance, merchantability or fitness for a particular purpose.

Because programs are inherently complex and may not be completely free of errors, you are advised to verify your work. **In no event will Addison-Wesley or Spectrum be liable for direct, indirect, special, incidental, or consequential damages arising out of the use of or inability to use the program, documentation or media,** even if advised of the possibility of such damages. Specifically, neither Addison-Wesley nor Spectrum is responsible for any costs including, but not limited to, those incurred as result of lost profits or revenue (in any case, the program must be used only for educational purposes, as required by your license), loss of use of the computer program, loss of data, the costs of recovering such programs or data, the cost of any substitute program, claims by third parties, or for other similar costs. In no case shall the liability of Addison-Wesley exceed the amount of the license fee and in no case shall Spectrum have any liability.

The warranty and remedies set forth above are exclusive and in lieu of all others, oral or written, express or implied. No Addison-Wesley dealer, distributor, agent, or employee is authorized to make any modification or addition to this warranty.

Some statutes do not allow the exclusion of implied warranties; if any implied warranties are found to exist, they are hereby limited in duration to the 90-day life of the express warranties given above. Some states do not allow the exclusion or limitation of incidental or consequential damages, nor any limitation on how long implied warranties last, so these limitations may not apply to you. This warranty gives you specific legal rights and you may also have other rights which vary from state to state.

To obtain performance of this warranty, return the item with dated proof of purchase within 90 days of the purchase date to: Addison-Wesley Publishing Company, Inc., Educational Software Division, Jacob Way, Reading, MA 01867.

The Student Edition of
MICRO-CAP III ™

An electronic circuit analysis program . . .
adapted for education

Martin S. Roden

California State University, Los Angeles

Addison-Wesley Publishing Company, Inc.
The Benjamin/Cummings Publishing Company, Inc.

Reading, Massachusetts • Redwood City, California • New York
Don Mills, Ontario • Wokingham, England • Amsterdam • Bonn • Sydney
Singapore • Tokyo • Madrid • San Juan • Milan • Paris

Addison-Wesley makes every effort to publish high-quality educational materials. If you have any comments regarding the contents of this manual, we would like to hear them. Please send them to: Addison-Wesley Publishing Company, Educational Software Division, Jacob Way, Reading MA, 01867.

The Student Edition of MICRO-CAP III is published by Addison-Wesley Publishing Company, Inc. and The Benjamin/Cummings Publishing Company, Inc. Contributors include:

Alan Jacobs, Executive Editor
Holly Wallace, Associate Project Manager
Karen Wernholm, Software Production Supervisor
Myrna D'Addario, Compositor and Text Designer
Jean Seal, Cover Designer

The Student Edition of MICRO-CAP III was developed and programmed by Spectrum Software, Inc.

This manual was prepared using Ventura Publisher on a Hewlett-Packard Series II printer.

MICRO-CAP III is a trademark of Spectrum Software, Inc. IBM and IBM PC are registered trademarks of International Business Machines Corporation. MS-DOS and Microsoft Mouse are trademarks of Microsoft Corporation. Hercules Graphics Card is a trademark of Hercules Computer Technology.

General Notice: Some of the product names used herein have been used for identification purposes only and may be trademarks of their respective companies.

ISBN   0-201-95414-1 manual

0-201-50605-X 5 1/4" package

0-201-50606-8 3 1/2" package

2 3 4 5 6 7 8 9 10-AL-9594939291

Preface

Welcome to the world of computer-based circuit analysis with MICRO-CAP III. The Student Edition of MICRO-CAP III is an analog circuit simulator based on numerical algorithms from SPICE2. MICRO-CAP III makes it exceptionally easy to enter your circuits into a personal computer because the entry process is menu-driven; the circuit schematic unfolds before your eyes as you interact directly with the software.

Circuit simulation programs are extremely useful in design. A real-life circuit does not behave exactly as the ideal theory predicts. Simplified models of elements and devices exclude many of the more complex dependencies that occur when the circuit is constructed. Computer simulation programs allow you to check the performance of a circuit before building it. It is easy to make changes in the circuit at this stage since you need change only the input to the computer program. This avoids a lot of time and expense and allows you to fine-tune your design for optimum performance. You can perform multiple simulations to examine worst case or find the probability that a component or system will fail

You have bought the student version of a software package used by engineering professionals. The graduating engineer cannot survive in the technological world without an understanding of, and familiarity with, the use of the computer as a design tool. Computer-aided engineering (CAE) is an integral part of electrical engineering. Designers use state-of-the-art computer systems to optimize component selection. Even small personal computers can aid in checking designs and indicating which parameters must be modified.

MICRO-CAP III is becoming the industry standard for use in electronic circuit design. Its SPICE-based circuit analysis program contains models of many popular electronic devices. This program forms an ideal base of a student-oriented electronic circuit-design program, not only because students will use the professional version after they graduate but also because the program's general features are representative of a broad class of analysis software.

If this is your first experience with simulation, a word of caution: Just as the proliferation of calculators did not eliminate the need to understand the theory of mathematics, electronic circuit simulation programs do not eliminate the need to understand electronic theory. As with calculators, this software can free the engineer from tedious calculations, permitting more time for doing the kind of creative work a computer cannot do.

Before performing a computer analysis of a network, you should have some idea of what to expect. We suggest you use this software to check your designs. In the process, you may uncover some unexpected results because a paper design rarely incorporates models as sophisticated as those used by this program.

Objectives

The primary objectives of the Student Edition of MICRO-CAP III are as follows:

- To provide a tool for handling the tedious calculations of circuit design, thus affording you more time for creative design work.

- To aid you in designing circuit boards for coursework and prepare you for using MICRO-CAP III in your profession.

- To provide a package that has been carefully designed to save you money while not compromising features significant at this phase in your education.

Features

The Student Edition of MICRO-CAP III includes the following features:

- A large library of standard passive and active devices, including popular models of BJTs, MOSFETs, JFETs, op-amps, and diodes.

- The capacity to custom-define devices and add them to the library for later use.

- Four types of analysis: Transient analysis, AC analysis, DC analysis, and Fourier analysis. Within each type of analysis, you can perform iterative analysis (temperature stepping and parameter stepping).

Organization of This Manual

The student manual is organized into three parts: Part I, "Getting Started," contains introductory information and instructions for entering the programs. It also contains a simple example so that you can experience the features of the program within minutes of setting it up on your computer. Part II, "Tutorials," is the heart of the manual. It contains examples and models, giving you a feel for the software's versatility and motivating a more in-depth approach as you progress. Each tutorial reinforces instruction with examples and problems. Part III, "Reference," contains three appendixes. The first is a glossary, containing a comprehensive alphabetical guide to program features. The second appendix contains additional information regarding the models used by MICRO-CAP III. The intent of this appendix is to draw a correlation between the symbols used in the models and the parameters found on standard data sheets. The third appendix is a quick reference listing of keyboard commands. This should be useful if you do not have a mouse.

Difference Between Student Edition and Professional Version

The Student Edition possesses almost all the features of the powerful professional version, with sufficient capacity for circuits that students might encounter. The elements of the Student Edition that differ from the professional version are as follows:

- Circuits can have a maximum of 30 nodes in the Student Edition.
- There is neither a Parameter Estimation Program (PEP) nor a plotter module.
- The program cannot perform Monte Carlo analysis.
- MESFETs and magnetic cores are not available.

Since the professional version formed the basis for the student version, some of the above functions do appear in the various menus, but they cannot be accessed in the analysis.

Acknowledgments

More than 30,000 copies of MICRO-CAP II, Student Edition, have been used by students in universities throughout the country. The manual and the software have therefore been thoroughly tested. We appreciate the response we have received, both solicited and unsolicited.

Sincere thanks to the following people for their valuable help during the development of this material:

Lt. Colonel Al Batten and his students, U.S. Air Force Academy, Colorado;
Henry Zmuda, Stevens Institute of Technology, Hoboken, New Jersey;
Edmond G. Case, Capitol College, Laurel, Maryland;
Donald H. Lenhert, Kansas State University, Manhattan, Kansas;
Glen C. Gerhard and his students at Arizona State University, Tempe, Arizona, where Dr. Gerhard was a visiting professor. He has now returned to the University of New Hampshire, Durham, New Hampshire.
Stanley Quon, a student at Cal State L.A., for thoroughly checking the manual and the program.

A product and manual of this complexity go through many iterations and extensive testing. Many people played key roles during this development process. The software was developed by Spectrum Software of Sunnyvale, California, where Andy Thompson and Don Chan spent long hours responding to perceived student needs and carefully reviewing all aspects of the project and the manual. The software was extensively tested by Paul Horvick of ISES (Independent Software Evaluation Service), Minneapolis, Minnesota. At Addison-Wesley, the following people worked long and hard to make this product as responsive and error free as possible: Holly Wallace (project manager), Karen Wernholm (production supervisor), and Myrna D'Addario (compositor). Without the efforts of these highly competent people, the project would not have successfully reached conclusion.

This manual was written by one of the authors of the text *Electronic Design—Circuits and Systems* by C. J. Savant Jr., Martin S. Roden, and Gordon L. Carpenter (Benjamin/Cummings, 1987, Second Edition, 1991). Although the Student Edition of MICRO-CAP III stands by itself, the wide variety of design approaches and examples in the textbook will serve as a valuable adjunct.

I am confident that you will find this package to be educational, exciting, and useful. We welcome your comments.

Martin S. Roden
California State University, Los Angeles

- To provide a tool for handling the tedious calculations of circuit design, thus affording you more time for creative design work.
- To aid you in designing circuit boards for coursework and prepare you for using MICRO-CAP III in your profession.
- To provide a package that has been carefully designed to save you money while not compromising features significant at this phase in your education.

Features

The Student Edition of MICRO-CAP III includes the following features:

- A large library of standard passive and active devices, including popular models of BJTs, MOSFETs, JFETs, op-amps, and diodes.
- The capacity to custom-define devices and add them to the library for later use.
- Four types of analysis: Transient analysis, AC analysis, DC analysis, and Fourier analysis. Within each type of analysis, you can perform iterative analysis (temperature stepping and parameter stepping).

Organization of This Manual

The student manual is organized into three parts: Part I, "Getting Started," contains introductory information and instructions for entering the programs. It also contains a simple example so that you can experience the features of the program within minutes of setting it up on your computer. Part II, "Tutorials," is the heart of the manual. It contains examples and models, giving you a feel for the software's versatility and motivating a more in-depth approach as you progress. Each tutorial reinforces instruction with examples and problems. Part III, "Reference," contains three appendixes. The first is a glossary, containing a comprehensive alphabetical guide to program features. The second appendix contains additional information regarding the models used by MICRO-CAP III. The intent of this appendix is to draw a correlation between the symbols used in the models and the parameters found on standard data sheets. The third appendix is a quick reference listing of keyboard commands. This should be useful if you do not have a mouse.

Difference Between Student Edition and Professional Version

The Student Edition possesses almost all the features of the powerful professional version, with sufficient capacity for circuits that students might encounter. The elements of the Student Edition that differ from the professional version are as follows:

- Circuits can have a maximum of 30 nodes in the Student Edition.
- There is neither a Parameter Estimation Program (PEP) nor a plotter module.
- The program cannot perform Monte Carlo analysis.
- MESFETs and magnetic cores are not available.

Since the professional version formed the basis for the student version, some of the above functions do appear in the various menus, but they cannot be accessed in the analysis.

Acknowledgments

More than 30,000 copies of MICRO-CAP II, Student Edition, have been used by students in universities throughout the country. The manual and the software have therefore been thoroughly tested. We appreciate the response we have received, both solicited and unsolicited.

Part III   Reference Section

1

GETTING STARTED

1

Before You Begin

In this chapter, we describe the contents of the Student Edition of the MICRO-CAP III package and the typographical conventions. You should carefully read this before you attempt to use or install the program.

Checking Your Computer Setup

System requirements for the Student Edition of MICRO-CAP III are:

- IBM PC, XT, AT, PS/2, or compatible
- MS-DOS or PC-DOS version 3.0 or higher
- At least 640K of RAM
- One of the following graphics adapters:

  IBM Color Graphics Adapter (CGA) and monitor

  IBM Enhanced Graphics Adapter (EGA) and color monitor

  IBM Multi-color Graphics Array (MCGA) and monitor

  IBM Video Graphics Array (VGA) and monitor

Hercules Adapter and monochrome monitor

- Two 360K disk drives or a hard disk (a hard disk is recommended)

- Mouse (the program can also be used without a mouse, but a mouse is strongly recommended)

Checking Your MICRO-CAP III Package

The Student Edition of MICRO-CAP III includes:

- Student User Manual

- Warranty registration card

- Two 5 1/4" 360K or one 3 1/2" 800K program disks containing all the following files :

Disk A—Program Disk

INSTALL.BAT

MC3S.EXE

MC#.DAT

Disk B—Data Disk

INSTALL.BAT

README.DOC

MSHERC.COM

Disk B—Subdirectory \SYS

BITMAP1.MC3

BITMAP2.MC3

BITMAP3.MC3

COMPEDIT.MC3

HELP.MC3

LOGO.MC3

RESOURCE.MC3

SHAPEDIT.MC3

Disk B—Subdirectory \DATA

CURVES.CIR

DIFFAMP.CIR

GUMMEL.CIR

L1.CIR

PRLC.CIR

SAMPLE.USR

STD.LBR

Disk A has only about 5000 bytes free, and Disk B has about 88,000 bytes free. If you are going to use the program on floppy disks, you need to be concerned with available space for storing your own data. Clearly Disk B has more space than Disk A. You may wish to delete some of the files to make more room on the disk. In particular, MSHERC.COM (6749 bytes) is used only if you have a Hercules graphics card. The .CIR files (about 1000 to 2000 bytes each) contain circuits that you can use as examples. You may wish to delete these. Once you read the README.DOC file, you can delete that, too.

Product Support

Neither Addison-Wesley nor Spectrum Software, Inc. provides telephone assistance to students for the Student Edition of MICRO-CAP III. However, phone assistance is provided to registered instructors who have adopted the Student Edition.

If you encounter difficulty using the Student Edition software, first refer to the Help screen or the sections of this manual that contain information on the commands or procedures you are trying to perform.

If you have to ask your instructor for assistance, describe your question or problem in detail. Write down what you were doing (the steps or procedures you followed) when the problem occurred. Also write down the exact error message (if any).

Typographical Conventions and Definitions

Keystrokes that you will type are designated in this manual by boldface type (for example, **PULSE**). Even though this manual often uses uppercase and lowercase letters, the program does not recognize the difference.

The following terms are used in this manual:

Type: The keystrokes to be typed are in a bold typeface; for example:

Type: **PULSE**

If several keystrokes are to be entered in sequence, such as

Type: **A**

Type: **PULSE**

then (to conserve space), the commands will often be shown on one line separated by semicolons. Do not type those semicolons when entering the commands; the semicolons are there just to symbolize a slight pause between entering each pair of commands; for example,

Type: **A; PULSE**

If keys are to be pressed simultaneously, we use a + sign. Thus, Shift+F2 means that you hold down the shift key while pressing F2.

Enter: The symbol for the Return (Enter) key is ⏎. When you see this symbol, press the Return key.

Click: Most mice have two or more buttons. The expression *click the mouse* means to press and release either the left or right button. The middle button, if present, is never used.

Click and drag: This is the process of positioning the mouse arrow in a selected location, pressing the left button, and moving the mouse while holding the button down. This is used to position components and to view and select from pull-down menus.

Cursor: This is the flashing object on your display. The mouse location is shown by the arrow. When you click the mouse, the cursor position moves to the mouse arrow location.

$\boxed{\text{KEYBOARD}}$ Although the mouse is preferred, keyboard instructions are provided; they are preceded by the heading to the left of this paragraph. The instructions for keyboard users follow the instructions for mouse users. Even mouse users may wish to alternate with use of the keyboard. The two input techniques can be interchanged at will. An appendix contains a quick reference to keyboard commands.

Select: Item selection is done with the mouse by moving the mouse arrow to the item and clicking the left button. It is done with the keyboard by pressing the Tab key and the up and down arrow keys until the flashing cursor is at the desired item and then pressing the space bar. Selected items are highlighted.

2

Initializing and Installing the Student Edition of MICRO-CAP III

In this chapter, we lead you through the steps you should follow before you can use the Student Edition of MICRO-CAP III: preparing disks, making backup copies of the original disks, and installing MICRO-CAP III so that it will work with your equipment. The procedures vary depending on the type of system you have.

You will need the following items:

- The disks that came with the MICRO-CAP III program (two 5 1/4" disks or one 3 1/2" disk).

- The DOS system disk or DOS installed on your hard disk.

- An IBM PC, XT, AT, PS/2 or a compatible system with two floppy disk drives or one floppy disk drive and a hard disk.

Working with Disks

Although we assume that you are already somewhat familiar with the operation of your computer, we begin by emphasizing some important points to keep in mind when you work with disks:

- Do not touch the exposed areas of the disk. If you are using 3½" disks, do not handle a disk with the shutter (the sliding metal door at the bottom center of the disk) open.

- If you are using 5¼" disks, take care when you write on the disk label; a sharp point or hard pressure may damage the disk. Use a felt-tip pen to write on a label that is already on the disk.

- Always place the 5¼" disk back in the sleeve after use.

- Keep the disk away from heat, sunlight, smoke, and magnetic fields (found around telephones, televisions, and transformers).

- Avoid removing a disk while the drive access light is on.

Starting Your Computer

Before you can prepare disks or make copies of your MICRO-CAP III disks, you must load DOS (the disk operating system), which lets the computer do basic tasks such as copying and formatting disks. Your computer should be off when you start this section.

1. (This step is for two-disk systems only; skip to step 2 if you are using a hard disk.) Insert your DOS disk in drive A.

2. Turn on the computer.

3. Enter the date (if necessary) in MM-DD-YY format, and press ↵.

4. Enter the time (if necessary) in HH:MM 24-hour format, and press ↵.

When you have finished entering the date and time, the operating system prompt appears. This manual uses A> for two-disk systems and C> for hard-disk systems. (Your prompt may look somewhat different.)

Installing the Program

Installation instructions differ depending on whether you have a hard disk or two floppy disk drives.

Installing on Floppy Disks

The two disks supplied are ready to run; therefore, you should simply copy them to create working disks. It is extremely important that you make copies of the MICRO-CAP III disks and use them as your working copies. Never use your original disks. If your working copy is lost or damaged, you still have your original disks from which to copy. To copy the disks, you should start with two formatted disks. The following instructions for formatting and copying disks relate to a system with two floppy disk drives. For other configurations, you must make appropriate modifications.

Formatting your disks: Now you can format your blank disks. To format the blank disk, with the DOS disk in drive A,

Type: **FORMAT B:** ↵

When the format is complete, DOS will ask **Format another (Y/N)?**

Type **Y** (and press ↵ in older versions of DOS) if you want to continue to format disks, or type **N** (and press ↵ if needed) if you are finished with your formatting.

CAUTION
Copying the disks: Before you begin, make sure that the original MICRO-CAP III disks are write-protected so that you cannot accidentally alter or erase any information on them.

1. Insert the Program Disk in drive A.

2. Insert a formatted blank disk in drive B.

3. Type **COPY *.* B:** ↵

4. Repeat the process for the Data Disk. In the case of this disk, you will have to create two subdirectories, \SYS and \DATA, and then copy the appropriate files to these subdirectories. Alternatively, you can use the DISKCOPY or the XCOPY command. See your DOS manual for details.

Installing on Hard Disks

Switch to drive A by typing **A:.** Your screen will now display the prompt A>. **Your current drive must be the drive your disk is in to install the program.** Place the original Program Disk into drive A.

Type: **INSTALL C:\MC3S**

We assume that your hard drive is drive C and that you wish to install in a directory named MC3S. You may alter these as needed. (In fact, we used a directory named MC3STUD in generating the figures in this manual since we already were using MC3S for an earlier version of the software.)

The INSTALL program will prompt you to place the appropriate disks in drive A. (After installation, the message on your screen may refer to the directory as MC3. This is still your MICRO-CAP III directory.)

Starting the Program

Before you run the main program, you may need to do one or both of the following:

1. If you are using a mouse, you must enable it. This is done by executing the MOUSE.COM program (or an equivalent routine from your mouse system software).

2. If you are using a Hercules graphics card, you must enable it. This is done by executing the MSHERC.COM program found on the MC3S main directory.

Since these two tasks need to be performed only once upon turning on your computer, you may wish to execute them as part of the AUTOEXEC.BAT file.

You are now ready to start the program. If you are using floppy disks, log on to the drive with the program disk. When you use your working copies of the disks, do *not* put a write protect tab on them, or you will not be able to run the program. Each time you use those disks, MICRO-CAP III updates the disks and therefore needs to write information to the disks in order to run. If you are using a hard disk, change to the MC3S subdirectory by typing **CD\MC3S**. Then

Type: **MC3S** ↵

If this is the first time running the program, the system may ask for the new data path and the new sys path. If you are using a hard disk, type **C:\MC3S\DATA** for the data path and **C:\MC3S\SYS** for the sys path.

If you are using floppy disks, type **B:\DATA** for the data path and **B:\SYS** for the sys path assuming that the data disk is in drive B. If it is in drive A, type **A** in place of **B**.

If you are installing from a 3 1/2" disk, you may still be prompted to put the Data disk in the drive, even though the data files are already there. Just press ↵.

You will learn how to change the data path for circuits and outputs in the section "File Menu" in Tutorial 1.

3

Introduction to the Student Edition

Now that you have installed the software, you are ready to explore the power of this simulation program. We begin our study by whetting your appetite with a simple example. You will be running a simulation within the next ten minutes.

An Easy Example

Although you have probably bought this program to analyze complex active electronic circuits, we illustrate its simplicity of use and great power with a simple passive circuit analysis. We analyze the RLC circuit shown in Figure 1, where the source is a pulse of height 5 volts and duration 400 nanoseconds (nsec), as indicated on the diagram.

We perform three types of analysis on this circuit. First, we find the output voltage, $v_o(t)$. Second, we plot the frequency response of the circuit (that is, the complex transfer function). Finally, we perform a DC analysis and plot the output voltage versus input voltage under DC steady-state conditions.

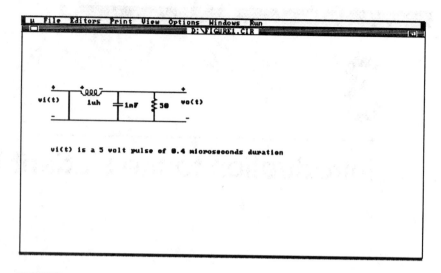

Figure 1

But first, we need to draw the circuit. MICRO-CAP III has one of the easiest interactive drawing systems of any circuit analysis program. Instead of having to lay out the entire circuit in advance and feed it into the computer using node numbers (as in SPICE), you actually draw the circuit on the monitor, component by component. As easy as this is, it still requires practice; you will become adept at it only by experimenting (and learning from your mistakes). The basics of element drawing can be learned in about five minutes, so take the time to work with the following example.

You should be sitting at your computer as you read the next few pages.

Drawing the Circuit

We assume that you have already executed MC3S. You are now looking at a screen with borders, pull-down menus at the top, and commands at the sides, as shown in Figure 2.

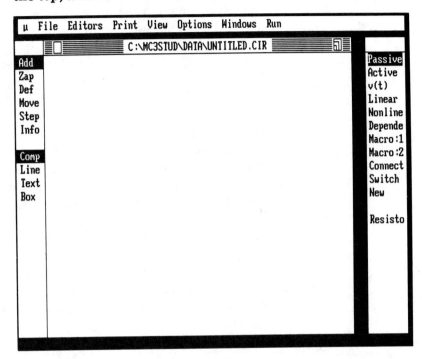

Figure 2

You can create circuits from this screen using either a mouse or the keyboard. We begin by giving instructions for the mouse. We follow with instructions for keyboard users, which are preceded by KEYBOARD .

We begin by drawing the pulse source at the left of the circuit. The highlighted commands on the left border indicate *Add* and *Comp*. Thus, the program is already in the mode of adding components to the diagram. If this were not so, we would have to invoke these commands by moving the curser to the appropriate command and clicking the mouse. Keyboard users will invoke these by pressing keys, as outlined below.

We wish to select a voltage source, which is denoted by $v(t)$ in the component window along the right border of this window. Select this component by moving the arrow to it and clicking the left button on the mouse. You will then be presented with a number of choices of source type. Select *Pulse* from this list by clicking the mouse. Below the menu, the screen will indicate that you have selected *Pulse*. You then need to add the component somewhere near the left edge of the drawing screen. Since this circuit is far smaller than the capacity of the program (our circuit has 2 nodes plus ground, and your program can handle up to 30 nodes), the specific location is not critical. Use the mouse to move the cursor to the appropriate position, and press the left button. If you hold down that button, you can drag the element anywhere on the screen. Then, while holding the left button, try clicking the right button. You will see the schematic figure rotate into any of eight positions (four 90-degree rotations and mirror images of each of these; in the case of simple two-terminal devices, the mirror images are identical to the original drawing, so there are only four distinct orientations). Once you are satisfied with the location of the source, release the buttons on the mouse.

KEYBOARD The cursor (a flashing solid square) is in the upper-left portion of the display. As you enter descriptions of each component, the program will draw the component starting at the cursor location. You can reposition the cursor by using the directional arrow keys on your keyboard. Try this a few times to get a feel for it.

Now locate the cursor somewhere in the left portion of the screen. Since the circuit we are entering is much smaller than the capacity of the drawing board, the starting point is not critical. On highly complex circuits, you will probably want to start at the appropriate edge.

The first component we enter is the voltage source.

Type: **A** for Add

and the system will prompt you for the name of the component to add. Now

Type: **PULSE** ↵

and the Pulse symbol will be displayed on the screen. Use the space bar to change the orientation and the cursor keys to change the location. When you are satisfied with the location and orientation, press ↵.

Now that the component is drawn on the screen (using either the mouse or the keyboard), the system will prompt you with a selection of available pulse source types from the library. Select *Pulse* either by clicking with the mouse or by using the cursor keys and pressing ↵.

Your screen should now look like Figure 3.

If you make any errors in entering a component, you can use the *Zap* command to remove any element or the *Def* command to change its value. We study this in Tutorial 1. For now, if you make a mistake, we suggest you clear the screen and start over. Do this by pulling down the File menu from

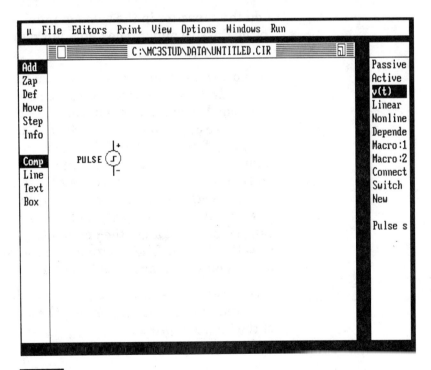

Figure 3

the top of the screen and clicking the *Create new circuit* entry. Alternatively, type **F** and then enter the menu number, or use the cursor and press ↵ to select from the menu. You will then be asked whether or not you wish to save the circuit. Click No with the mouse, or type **N**. The screen will then clear.

We are ready to add the inductor. Inductors are part of the Passive menu, so select that menu either by clicking the mouse on top of *Passive* and then clicking *Inductor* or by typing **A; Inductor.**

You are now ready to position the inductor as you did the source already entered. Once the element is drawn, you must enter the value. We want this to be 1 microhenry, so enter 1E-6. Alternatively, you can enter 1U or 1UH. MICRO-CAP III recognizes these as equivalent statements. It makes no difference whether you enter these as uppercase or lowercase letters. The program interprets everything as uppercase.

Add the capacitor in the same manner. Capacitors are found in the Passive menu. The value is 1 nanofarad, which is entered as 1E-9 or 1NF or 1N.

We now want to enter the horizontal line at the bottom of the circuit. Note that the left side of the screen contains a *Line* instruction below the *Comp* heading we have been using. Place the mouse cursor on this and click. You are now in the mode of adding lines, and the right side menu is disabled. You add a line by specifying the endpoints. Place the cursor at one end of the line, and press the left button. Then move the cursor to the other end, and press the right button. Since the line drawn by the program must be either horizontal or vertical, you need not align the endpoints perfectly; the program does this for you. The program also contains a feature that aligns points that are close together.

KEYBOARD First position the cursor at one end of the line. Type **L** to invoke the line command. Position the cursor at the other end of the line, and press ↵.

Use this technique to add both the bottom line and the horizontal line at the upper right of the circuit. The screen should now look like Figure 4.

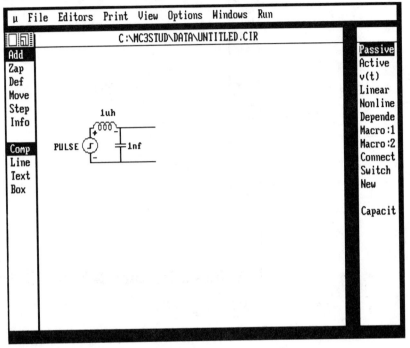

Figure 4

We now wish to add the resistor at the right side. Do this from the *Passive* element menu, and set the value to 50.

Congratulations! You have now entered the circuit, and your screen should look like Figure 5.

You should get into the habit of saving circuits. We discuss this in Tutorial 1. Pull down the File menu (or type **F**), select *Save circuit as* and type **C:\MC3S\DATA\Example.** This saves the circuit in a file named EXAMPLE.CIR.

If you tried to analyze the circuit of Figure 5, you would receive an error message. The reason is that all circuits *must include a ground symbol* connected to some component somewhere in the circuit. The ground symbol appears within the

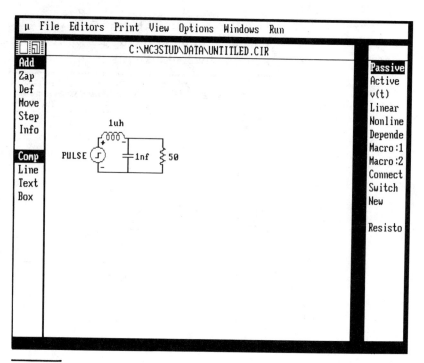

Figure 5

Connector group. Once it is selected, we add it in the same manner as all other components. Mouse users position the arrow at the appropriate point and click the left button.

KEYBOARD Keyboard users type **A; Ground**.

The circuit is now complete! You can analyze this circuit without storing it on the disk. On the other hand, if you want to refer to it later, you could store it prior to running the analysis.

The analysis yields parameters at various points in the circuit. We need some way of identifying these points. MICRO-CAP III automatically assigns a number to each node in the circuit. You can display these node assignments either by pulling down the View menu and selecting *show node numbers* (position the mouse arrow over the View menu and pull down that menu; select *show node numbers* by moving the mouse downward while continuing to hold the left button, until the correct entry is highlighted, and then releasing the button) or by typing **V** and then **7**. The screen will then appear as in Figure 6.

For our circuit, the input voltage appears at node 1 and the output at node 2, where both voltages are measured relative to ground.

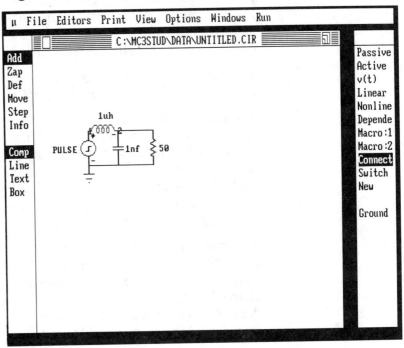

Figure 6

Transient Analysis

We first perform a transient analysis. That is, we plot the output voltage as a function of time with the input pulse as specified. The analysis is initiated from the Run menu that appears at the top of the screen. Using the mouse, position the arrow over the Run Menu and pull down that menu. It includes four entries. Select *Transient* by moving the mouse downward (while continuing to hold the left button) until the correct entry is highlighted, and then release the button.

KEYBOARD   The transient analysis is initiated using the keyboard by typing **R; 1.**

Once the transient analysis is initiated, you will be presented with a Limits window (If you have not previously titled the circuit, the program will ask you to supply a name.) We discuss this window in detail in Tutorial 2. The window sets the various analysis limits. It also includes specification of which parameters should be plotted and the ranges for these plots. For new circuits, the system automatically provides a set of default limits and picks the voltages at nodes 1 and 2 for plotting. The default values are acceptable for this example.

The analysis is initiated by pulling down the Transient menu and selecting *Run* or by pressing F2.

The program then runs the simulation from t = 0 to t = 1 microsecond, and displays the voltage waveforms at nodes 1 and 2. Node 1 represents the input, and the program displays that voltage on axes with a range from 0 volt to 5 volts. Node 2 represents the output, and the program displays that voltage on axes with a range from –1 volt to 6 volts. The resulting plots are shown in Figure 7.

The source generates a pulse close to 5 volts in amplitude and 400 nanoseconds in duration. The reason that the pulse in the figure is not exactly at this value is that the practical source has internal resistance with an associated voltage drop. The output attempts to follow the input but contains a transient and damped oscillations.

We hope you are now excited about your purchase. Evaluating the time function for the transient response would not be much fun without the computer. You would have to use either Fourier or Laplace analysis and then carefully find the inverse transform of the result or convolve two time functions together. And if the elements were to deviate from ideal, the "paper solution" of this circuit would become virtually intractable.

We exit the transient analysis either by pulling down the Run menu and selecting *Quit analysis* or by pressing F3.

AC Analysis

In AC analysis, we plot output versus input as a function of frequency. The straight line approximation to this plot is the

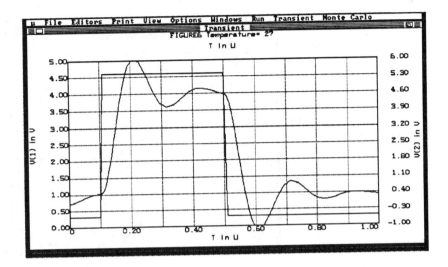

Figure 7

familiar bode plot. The AC analysis is run in a manner similar to that of the transient analysis. We have already drawn the circuit, and there is no need to make any changes. We therefore simply initiate the analysis either by using the mouse to pull down the Run menu to the *AC* entry or by typing **R; 2**.

The source is replaced by a 0.001-ohm resistor for the AC analysis, and we assume sinusoidal inputs and steady-state operation.

Initiation of the analysis presents a Limit menu. We again choose the default values, and initiate the simulation either by pulling down the AC menu and selecting *Run* or by pressing F2. If the circuit has not yet been saved, you are prompted to do so. The analysis then proceeds and generates curves as shown in Figure 8.

There are three curves. These represent gain, phase shift, and group delay. If you have trouble distinguishing between the three curves, don't worry. You will learn in Tutorial 2 how to add tokens to the curves so that they are easy to associate with the correct axis (even without color). The program analyzes frequencies from 1 MHz to 100 MHz and

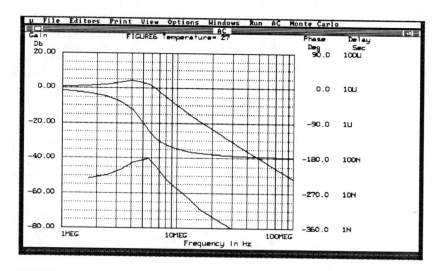

Figure 8

displays gain from –80 dB to 20 dB, phase shift from –360 degrees to +90 degrees, and group delay from 1 nsec to 0.1 msec. The input and output voltages are the same as specified in the transient analysis.

The amplitude plot starts at 0 dB for low frequency and approaches approximately –52 dB. The final slope of the amplitude curve is –12 dB/octave, as the theory predicts.

The phase of the transfer function starts at approximately 0 degrees and ends up at approximately –180 degrees for high frequency.

You will probably not be using group delay in your basic electronics course. It relates to the envelope distortion produced by a system. When the phase curve is linear, there is no distortion because all frequency components of the input are delayed by the same amount of time. This time delay relates to the slope of the phase characteristic.

Once again, we exit the analysis either by pulling down the Run menu and selecting *Quit analysis* or by pressing F3.

DC Analysis

In DC analysis, we plot output versus input under DC conditions. Ideal capacitors would become open circuits, and ideal inductors would become short circuits. The DC analysis is run in the same manner as was the transient and AC analysis. We have already drawn the circuit, and there is no need to make any changes, although the analysis will replace the pulse input by a DC source. We initiate the analysis either by using the mouse to pull down the Run menu to the *DC* entry or by typing **R; 3**.

If the circuit has not yet been saved, you are prompted to do so. Initiation of the analysis presents a Limit menu. We again choose the default values and initiate the simulation either by pulling down the DC menu and selecting *Run* or by pressing F2. The analysis generates curves as shown in Figure 9.

The input/output characteristic is a straight line of unity slope. This finding is not surprising since, under DC conditions, the capacitor is an open circuit, and the inductor is a short circuit. Thus, the output voltage will equal the input voltage. The DC analysis mode becomes more interesting when nonlinear devices are present in the circuit.

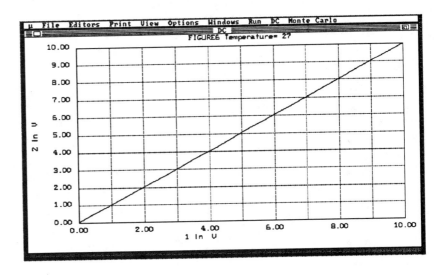

Figure 9

When you have completed this, you can exit the analysis either by pulling down the Run menu and selecting *Quit analysis* or by pressing F3.

We have illustrated the three circuit analysis options for a simple example. Even though the circuit we analyzed did not contain any nonlinear passive devices or any active electronic devices, we hope you can sense the excitement and tremendous power of this program.

In addition to transient, AC, and DC analysis, MICRO-CAP III is capable of doing Fourier analysis. In this mode, the program can plot the Fourier transform of a time function. You learn how to do this in Tutorial 5.

In Part Two, we take a more organized approach toward gaining skill with the program. We first learn how to draw more complex circuits. We then explore the four analysis packages in detail. Tutorials 2, 3, 4, and 5 are intended to stand on their own, so you may cover them in any order you wish.

II

TUTORIALS

In Part I of this manual, we presented an example to whet your appetite. We hope you found that RLC circuit both interesting and motivating. Although it introduced you to many features of this powerful software, it did so in the context of solving only one simple circuit.

This next part of the manual contains a detailed presentation of the major features of this program. There are several ways to learn to use this software. One way is to read all the tutorials and then start doing examples. However, if you are like most students, you will have forgotten the earlier portions of the chapter by the time you reach the end. A second way is to learn a few features at a time and practice with these until you are comfortable with them. Yet another way is to start using the software in your coursework, and learn the various levels of sophistication as you need them. This method proves best for most students. Many of the more complex features of the program may be of limited use to you, so the only purpose in reading about them now is to be aware that they are available if you need them later.

We would be unrealistic if we did not mention another very different way to learn this software. The software is quite friendly, and the menus are self-explanatory. It is therefore possible to use the software without first reading the manual. You would then resort to the manual only when you get into trouble. There is no way to damage your computer by misusing the software, and as long as you have made a backup copy, you need not worry about damaging the software. Go ahead, be reckless. Learning by mistake is a proven effective technique.

To close the loop and receive feedback in your learning attempts, we recommend that you solve the problems at the back of each tutorial. Your instructor has the answers to the problems.

1

Schematic Editor

Introduction

When you run MICRO-CAP III, you will immediately enter the editing mode, and your screen will look like Figure 10.

In this tutorial, we describe every entry on that screen, with the exception of the Run pull-down menu. We discuss the various Run options in Tutorials 2 through 5.

The Mode menu and the Components menu run down the left and right side of the screen, respectively. If you click and drag down the Windows menu from the top, you will find that there is a check mark next to each of these two windows. If you erase a window at any time and want it back, you need simply pull down the Windows menu and click the appropriate entry.

We discuss each menu entry in the order in which you may wish to use it, beginning with the most common. We list all menu entries below. The list contains the entire name, whereas the section headings that follow are abbreviated in the same way they are displayed on the monitor.

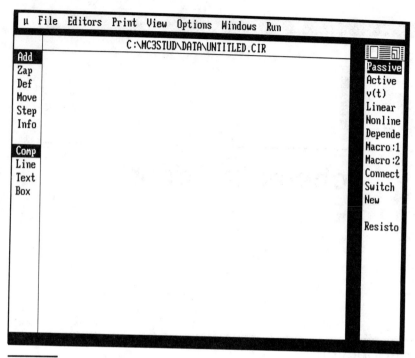

Figure 10

Add

    Components

        Passive Components

        Active Components

        Connections

        v(t) Sources

        Linear Sources

        Nonlinear Sources

        Dependent Sources

        Macro:1

        Macro:2

        Switches

Add

This mode is used to add an element to the existing circuit. The selected component will be added at the mouse arrow location when the mouse button is clicked. If using the keyboard, the component is added at the cursor location.

Comp

This allows you to choose any component specified in the component library in the right-hand menu.

When the *Comp* mode is selected, the Components menu along the right-hand border is activated. We describe each entry on that window below.

Drawing elements: Let us assume that you have selected the *Add/Comp* mode and the desired element from the Components menu. You then add the component as follows:

Move the mouse arrow to the desired location on the screen, and click the left button. If you hold that button down, you can drag the component to any location on the screen. If you click the right button while holding down the left button, you reorient the element into any of four rotations (90 degrees each) and the reflected (mirror image) versions of these.

You should take the time now to try this procedure. Assuming you have not used the program and selected a different element, your screen has probably highlighted passive components and the resistor. Therefore, if you click the mouse on the drawing screen, you will draw a resistor. Try moving it and rotating it. After releasing the button(s), the element is drawn on the screen. You must then type in the name or value of the element. If you again click the button, the element will be erased.

KEYBOARD

Type **C** (for Component) if it is not already highlighted in the Mode window. Then type **A** (for Add) followed by the name of the component. (Actually the first few letters of the component are enough, provided this gives a unique description; for example, **i** will lead to a current source, whereas **in** will add an inductor.) The element will then be added at the cursor location. You can move the element using the cursor arrow keys, and you can reorient it by pressing the space bar. Once it is positioned where you want it, press ↵ to fix the location and orientation. If you then press **ESC**, the element will be erased.

Passive

If you click the *Passive* Component menu entry, you are presented with a new menu of passive devices. This is illustrated in Figure 11.

You then click on the specific component you wish to draw. These include resistors, diodes, transformers, capacitors, inductors, and lines. (Cores are not supported in the Student

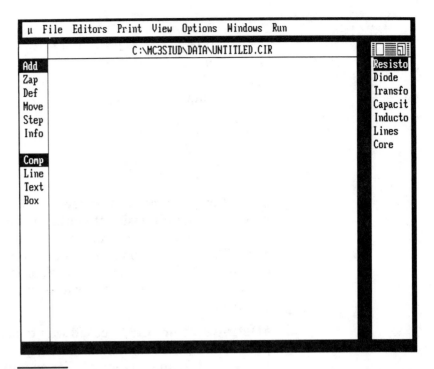

Figure 11

Edition, although they appear in the menu.) Once you select any of these elements and draw it by clicking the mouse, you will be prompted to enter the value. In the case of resistors, capacitors, and inductors, you enter the number of ohms, farads, or henries using the keyboard. You can enter numbers in any of three separate formats:

Real Numbers: You enter the value of the component. For example, 1 megohm would be 1000000, and 1 microfarad would be .000001.

Floating point (scientific notation): You enter using powers of 10. For example, 1 megohm would be 1E6, and 1 microfarad would be 1E-6. MICRO-CAP III recognizes no difference between uppercase and lowercase.

Engineering notation: You can use the following abbreviations for powers of 10:

F	Femto	1E-15
P	Pico	1E-12
N	Nano	1E-9
U	Micro	1E-6
M	Milli	1E-3
K	Kilo	1E3
MEG	Mega	1E6
G	Giga	1E9
T	Tera	1E12

Thus, 1 megohm could be entered as 1MEG or 1000K. You may also add unit designations after the abbreviation without affecting the value. Do not, however, use the F by itself; 1F is 10^{-15} farads rather than 1 farad. And 1 microfarad can be entered as 1UF, or 1000NF, or the F can be dropped. Do not leave any space between the numeric and the abbreviation.

#Define statement: A second way to define the value of a passive component is to give the component a name and then to use the #Define statement. For example, when prompted for the resistor value, you could use a name, for example, R1. Then, in any blank part of the diagram, you would enter text (we learn to do this in the section "Text" later in this tutorial) for the define statement in the form,

#Define R1 1200

Within this definition, you can use a number of mathematical operations and symbols (see the section "v(t) Sources"). For example,

#Define R1 50*(T<700ns)

would define a resistor of 50 ohms as long as the time is less than 700 nanoseconds. Following that time, the resistor value would drop to zero.

If you select *Diodes* and then add this component to your drawing, you will be asked to enter the diode designation from the library of diodes. The library contains 39 standard diodes and 6 user-definable diodes. You select the diode number by clicking the menu entry. We see how to modify the library in the section "Editors Menu."

KEYBOARD After drawing the diode (that is, position with cursors, rotate with space bar, and press ⏎ to draw), you can scroll through the library of diodes using the cursor arrow keys, and then press ⏎ to select a particular diode.

#Model statement: Another way to specify the diode being used in the circuit is to use the #Model statement. When prompted for the type of diode, type any symbol you choose, for example, D1. Then, in any blank space somewhere on the diagram, insert the model description text in the form,

#Model D1 D(IS=1P CJO=1P TT=1N TOL=50%)

We specify four parameters in this model statement. We discuss these parameters in the section "Editors Menu."

If you click the *Line* entry on the Passive component menu, you will add a transmission line to the diagram. This can be used only in transient analysis. You must specify the number of sections, the time delay for each step, and the characteristic impedance of the line.

If you select *Transformer*, you will have to define the primary inductance, the secondary inductance, and the coefficient of coupling. Thus, when prompted to type the value, you enter **LP,LS,K** for the three parameters; for example, you might enter **.01,.0001,.98**.

Active Components

When you select this item from the Components menu, you are presented with a second menu as shown in Figure 12.

Op-amps and transistors are included in this menu item. You will be presented with a list containing *op-amps, NPN* and *PNP* bipolar, *PMOS* and *DPMOS*, *NMOS* and *DNMOS*, and *NJFET* and *PJFET* transistors. The menu contains *MESFET*s although these are not supported in the student version of MICRO-CAP III.

Clicking any of these active devices presents a menu of those in the library. The libraries contain the following active

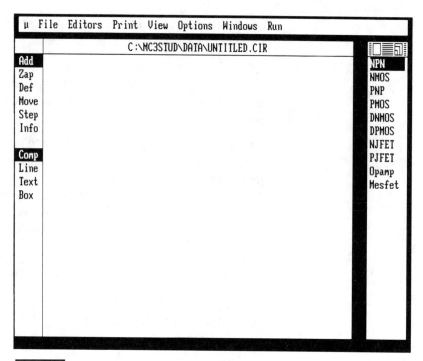

Figure 12

devices. You can add other devices to the library as discussed in the section "Editors Menu."

NPN and **PNP transistors:** Once you add either of these bipolar junction transistors to your circuit, you are presented with a library of 39 standard transistors and 10 additional user-defined transistors. Select the desired transistor by clicking the mouse on that entry. You can also use a #Model statement as we did for diodes.

NMOS, DNMOS, PMOS, and DPMOS: The D notation indicates discrete, and the absence of a D indicates an integrated circuit MOS transistor. The appropriate symbols appear on the screen when you add any of these active devices to your circuit. You are presented with a menu of 41 standard devices and ten user-definable devices.

NJFET and PJFET: When you add either type of JFET, you are presented with a menu of 49 standard devices and two user-definable entries.

OP-AMPS: When you add an op-amp, you are presented with a menu of 43 standard op-amp numbers, and two user-definable entries.

Connections

When you click the *Connections* entry in the Components menu, you are presented with another menu with 16 entries, as shown in Figure 13.

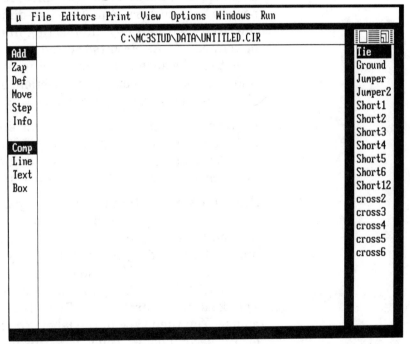

Figure 13

These fall into five categories: *Ground*, *Short*, *Jumper*, *Tie* and *Cross*.

Every circuit you draw on the screen must have a ground, otherwise attempts at analysis result in an error statement. Once you select the Ground, you add it to the circuit in the usual way (click the mouse, click and drag to move, click right button to rotate).

In a later section ("Line"), we discuss the easiest way to draw lines on the circuit. However, you sometimes need to draw short lines in order to separate components from each other (that is, add leads to a component). You can use the Connections menu to select any one of seven possible Shorts, where the number indicates the length of the short in number of grids.

When two lines cross in a drawing, it is assumed that a connection exists between the two lines. It would therefore appear to be necessary to make a two-dimensional drawing of the circuit without any intersecting lines (unless they are to be connected). As circuits become more complex, it is not possible to make such a planar drawing. We need ways to cross lines without an electrical connection. The *Tie*, *Jumper*, and *Cross* menu selections permit this.

A Cross or Jumper is used whenever lines that are not electrically connected intersect. You add this component and position the "loop" to lie over the second line. Two lengths of jumpers and five lengths of cross are available.

Sometimes points that are widely separated must be electrically connected. We could do this by drawing lines connecting the points, using jumpers where necessary. However, this may clutter the diagram unnecessarily. The Tie provides a simple alternative. You use this to mark the two points with any identical label. They are then electrically tied together. You position the arrow over one of the points and click the mouse. You then type in the label. You repeat the operation for the second point. You can do this for any number of points. An example is shown in Figure 14, where we have used both a jumper and a tie.

v(t) Sources

When you select *v(t)* from the Components menu, you are presented with six choices from a source menu, as shown in

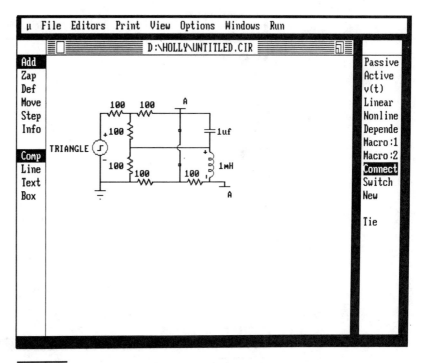

Figure 14

Figure 15. These are *Battery, Pulse Source, Sine Source, ISource, User Source,* and *User Function.*

If you select *Battery* and enter this on the diagram, you are prompted for the battery voltage. You can enter this from the keyboard or give the battery a symbol and specify its value with a #Define statement.

If you select *Pulse Source* and enter this on your circuit, you are presented with a menu of 21 possible pulse sources. These include the square pulse that we used in the simple example of the previous chapter. We learn how to define these sources in the section "Editors Menu."

If you select *Sine Source*, you are presented with a menu of 16 possible sine sources.

Isource is a DC current source. You specify the value after drawing the source on the screen in the standard manner.

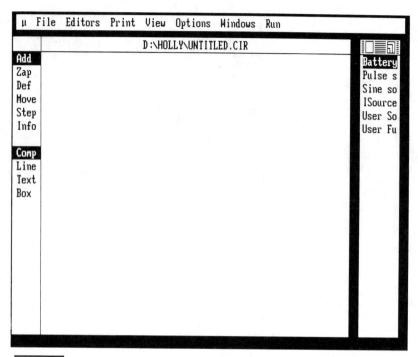

Figure 15

The *User Source* is a source that gets its values from a user-defined file in an ASCII text file. This file contains N sequential values representing the waveform at N successive time points. There are a number of ways of creating such a file, and we discuss this in Tutorial 5 when we examine Fourier analysis.

The *User Function* is a source defined by an algebraic expression. The expression may contain any valid terms, and is usually a time-dependent function. We could accomplish the same thing by using the nonlinear source described in the section "Nonlinear Sources." The User Function source is included primarily because it was part of MICRO-CAP II. Since MICRO-CAP III contains the more powerful nonlinear source, we recommend that you not use the User Function source. It may be dropped from later versions of MICRO-CAP.

Functions and equations: The previous two sources plus the dependent sources that we discuss in the next two sections can be defined by functions. Functions can also be used as part of the #Define statement for elements. We summarize the available functions below:

P(X,Y,Z)	Power flowing into a circuit section (X and Y define the nodes between which must exist a resistor or inductor that measures the current. The current is multiplied by the voltage drop between nodes Y and Z.)
E(X,Y,Z)	Energy flowing into a circuit section
T	Transient analysis simulation time
F	Real AC analysis frequency value
S	Complex radian frequency = $2\pi Fj$
+	Addition
−	Subtraction
*	Multiplication
/	Division
MOD	Modulus
SIN(x)	Sine function (x in radians)
COS(x)	Cosine function (x in radians)
TAN(x)	Tangent function (x in radians)
ATN(x)	Arc tangent function
SINH(x)	Hyperbolic sine
COSH(x)	Hyperbolic cosine
TANH(x)	Hyperbolic tangent
COTH(x)	Hyperbolic cotangent
LN(x)	Natural log
LOG(x)	Base 10 log
EXP(x)	e^x
ABS(x)	Absolute value
DB(x)	Decibels
SQRT(x)	Square root
SGN(x)	+1 if x > 0, −1 if x < 0, −1 if x = 0
^	Exponentiation operator
NOT	Negation operator
AND	AND operator
OR	OR operator
XOR	Exclusive-OR operator

Relational operators (return 1.0 if true, 0.0 if false):

<	Less than
>	Greater than
<=	Less than or equal to
>=	Greater than or equal to
<>	Not equal to
=	Equal to

Linear Sources

Linear sources are specified by Laplace transfer functions. When you select this type of source, you are presented with a menu of eight possibilities, as shown in Figure 16.

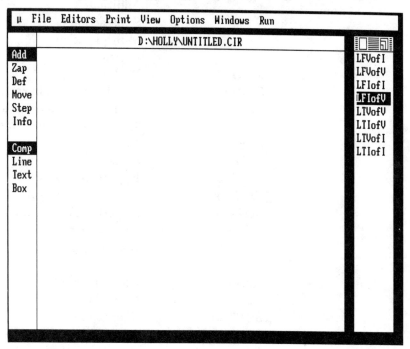

Figure 16

These represent controlled sources where the source and the controlling variable can be either voltages or currents. This yields four possible combinations (for example, voltage-controlled voltage source, current-controlled voltage source). Then, for each of these four possibilities, there are two types of sources: those specified by a formula and those specified by a table. The formula sources contain an F in the specification, and the tabular sources contain a T.

The formula sources can be used to simulate a transfer function. For example, if you specify the formula (after adding the source to the diagram) as $1/(1+.001*S)$, your controlled source is acting as a first-order, low-pass filter with 3-dB cutoff frequency at 1000 rps. Figure 17 shows the circuit for this function and the bode plot that results. The bode plot was produced using the AC analysis program, which we describe in Tutorial 3.

If you select a tabular source, you must specify that source in the form of a table. The table must contain ordered triplets of numbers: frequency, magnitude, and phase. You would specify the source with a list of the form,

(F1,M1,P1) (F2,M2,P2) .... (FN,MN,PN)

The frequency is in Hz. The sets of parentheses must be separated by spaces, and the frequencies must be in ascending order. The program interpolates between points, and for all frequencies below F1, the values are set at M1 and P1. For frequencies above FN, the values are constant at MN and PN. You enter this table either directly when prompted for the parameter value (after adding the source to your figure) or by use of a #Define statement.

Nonlinear Sources

When you select nonlinear sources, you are presented with a menu of eight possibilities, as shown in Figure 18.

These represent controlled sources where the source and the controlling variable can be either voltages or currents. This yields four possible combinations (for example, voltage-controlled voltage source, current-controlled voltage source). Then, for each of these four possibilities, there are two types of sources: those specified by a formula and those specified by

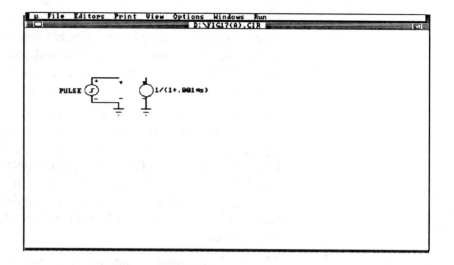

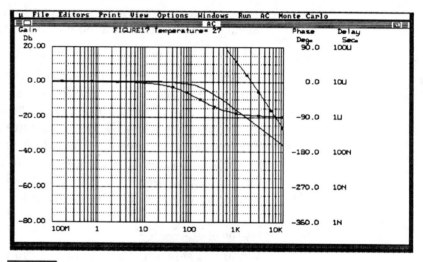

Figure 17

a table. The formula sources contain an F in the specification, and the tabular sources contain a T.

If, for example, you select NFVofI, you are selecting a current-controlled voltage source that is specified by a formula. The output variable is computed from an expression involving any of the following variables:

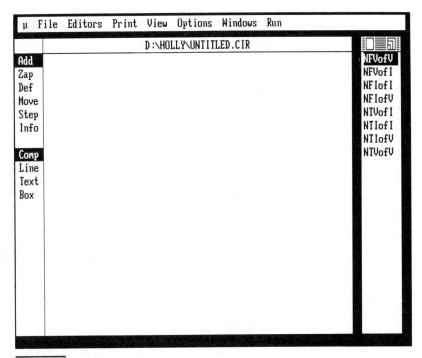

Figure 18

T	Time
VIN	Voltage across the input pins
V(A,B)	Voltage from node B to node A
TEMP	Ambient temperature
VT	Thermal voltage
π	3.14159. You can type **PI** for this constant
I(X)	Current through any resistor or inductor named X

You can also use the various mathematical expressions defined in the section "v(t) Sources." Figure 19 shows an example of a voltage-controlled voltage source. This is actually a VCO (voltage-controlled oscillator), where the input voltage controls the output frequency.

The output is a frequency modulated waveform whose frequency is linearly related to the input voltage.

Tabular nonlinear function sources are specified by a table of input values of the form,

(X1,Y1) (X2,Y2) ... (XN,YN)

The same rules apply as in the case of Linear sources.

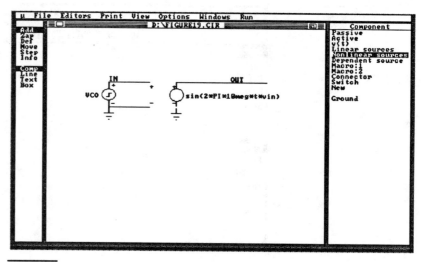

Figure 19

Dependent Sources

When you select *Dependent source* from the Components menu, you are presented with six choices, as shown in Figure 20.

Four of these are conventional dependent sources. If you add any of these to the diagram, you are prompted to enter the proportionality constant for that source. For example, if you select *VofV* and enter 5, the controlled source voltage is five times the voltage at the input nodes.

Polynomial V and I sources follow an input/output relationship as follows:

$$Y(X) = A + B*X^C + D*X^E + F*X^G$$

X is the input. In the case of a controlled source, it can be a voltage or current at some other point in the circuit. You can accomplish the same thing using nonlinear controlled sources. The polynomial source is included primarily because it appeared in MICRO-CAP II. Because the nonlinear source can be used in a much wider array of applications, the polynomial source may be removed from later versions of MICRO-CAP.

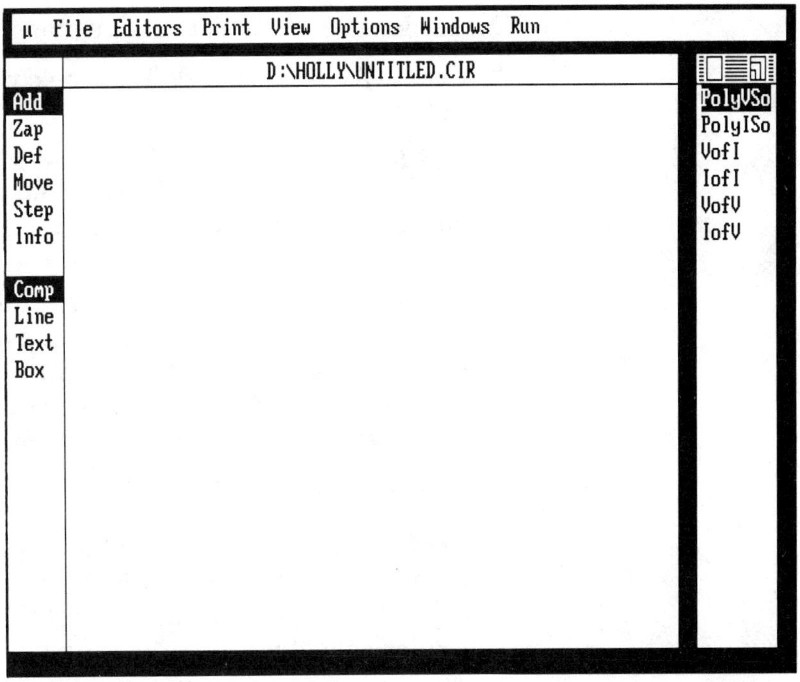

Figure 20

Macro:1

A macro is a circuit containing components, text, and lines. You use macros when a particular circuit configuration is repeatedly used in a larger network. When you select *Macro:1* from the Components menu, you are presented with a list of macros that are part of the professional version of MICRO-CAP III. To allow running the student version on computers without a hard disk, we have not included any macros on your disk. If you add a macro to a diagram, you are actually adding the expanded circuit. Since the simulation uses the expanded circuit, you must be careful not to exceed the 30-node limit of the Student Edition.

Provided that you have sufficient memory available, you may define other macros. This is useful if you incorporate a particular circuit configuration at many points within your designs.

Defining a macro consists of five steps:

1. Create a circuit that defines the contents you wish to be in the macro.

2. Add grid text labels at the circuit points that connect with the outer circuit. Therefore, if your macro has three leads, you need to define these three points using #PinN, where N varies from 1 to 3.

3. Add a #Parameter statement to indicate any parameters that the main circuit should request. For example, if your macro contains a resistor, R1, and a capacitor, C2, and you wish to specify these in your main circuit, you need a statement in the macro of the form,

   #Parameter(R1,C2)

   Then the name of the macro in the main circuit would be NAME(R1,C2), where NAME is the name you have assigned to the macro, and R1 and C2 are the appropriate numerical values.

4. Save the circuit in a disk file.

5. Add a new entry in the component library by selecting *New* from the window. You will have to specify:

   The macro name (the same name used to save it to disk)

   The name of the shape to be used when drawing the macro (you can pick any shape stored in the program)

   A definition type of MACRO

   Parameter text locations

   Pin assignments

At this point, the macro is available for use in another circuit. To use it, select the macro name from the Component window as usual, and place it in your circuit. When the parameter string is requested, enter the macro name and then, if needed, a list of parameters in parentheses.

As an example, suppose you wish to include an integrator into your circuits at many points. You decide to create a macro to do this. The macro could be as shown in Figure 21.

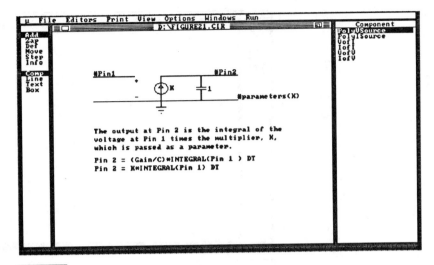

Figure 21

We have used a dependent voltage-controlled current source with proportionality factor K. This feeds a 1-farad capacitor, so the output voltage is K times the integral of the input voltage. We have labeled the two leads #Pin1 and #Pin2, and included a #kparameters(K) statement to indicate that the proportionality factor is included in the macro name in the main circuit.

Macro:2

Macro:2 is the same as Macro:1. Your new macro will be stored in this directory entry.

Switches

When you draw a switch on the circuit, you will note that the symbol includes a switch and two controlling nodes, as shown in Figure 22.

There are three types of switches that you can use in circuits: current-controlled, voltage-controlled, and time-controlled. If you use current-sensing, you must insert an inductor across the input nodes of the switch. Voltage-controlled switches are controlled by the voltage across the two input nodes, whereas

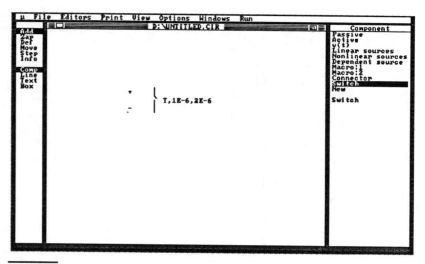

Figure 22

time-dependent switches use the time variable. In all cases, the open switch is represented by a high resistance, ROFF, and the closed switch is represented by a low resistance, RON. These resistance values can be set from the Options menu (see the section "Options Menu").

Switches are used only in transient analysis. After you draw a switch, you are prompted to type in the value. This is done in the following format:

S,N1,N2

S is the controlling parameter. Type **I** for current-controlled switches, **V** for voltage-controlled, and **T** for time-controlled. N1 and N2 are the threshold values of the controlled parameter between which the switch changes state. You can simulate either a normally open or normally closed switch depending on whether N1>N2 or N2>N1. If N2>N1, the switch is normally open, and it closes when the parameter is between N1 and N2. If N1>N2, the switch is normally closed, and it opens when the parameter is between N1 and N2.

New

This entry is available for user-defined components or macros. You can add these to the library.

Line

This mode allows you to draw straight lines. You do this by clicking the left button at one end and right button at the other.

`KEYBOARD` Move the cursor to one end of the line and type **L**. This initiates the line at the current cursor position. Use the arrow keys to move the cursor to the other end of the line, and press ↵ to terminate the line.

Text

Text can be placed at any of the grid locations using this menu selection. Move the mouse arrow to the desired text location, and click the left button. Enter the text, and then press ↵.

`KEYBOARD` Move the cursor to the desired position and type **T** followed by **A**. Enter the text, and then press ↵.

Box

This mode allows you to define a box for repetition (step) or Deletion (Zap). To define a box, select the Mode window entries *Def* and *Box*. Then click the left button to mark the upper-left corner and the right button to mark the lower-right corner.

`KEYBOARD` Move the cursor to the upper-left corner of the desired box and type **D**. Then move the cursor to the lower-right corner of the box and type **T** to terminate the definition.

Once the box is defined, you can copy the contents to another place on the circuit by using the Add and Step modes. You can delete the contents of the box with the Zap mode (see the section "Zap"), or move these contents to a new location using the Move mode (see the section "Move").

Zap

This selection is used for deleting elements, text, or the contents of a box. It operates in much the same way as the *Add* selection, but erases instead of draws.

Def

This selection has two distinctly different functions depending on whether you are using Boxes. It defines the box boundaries when in this mode. When not in this mode, it is used to change a component parameter or to change text.

Move

This selection is used to move an element. To move an element, the user clicks the element and drags it to the desired location. While the left button is held, the right button can be used to rotate the element, just as in the case of Adding elements.

KEYBOARD   To move an element using the keyboard, type **M** for Move. Then use the cursor keys to move the element to the desired location. Finish by pressing ↵.

Step

In drawing a complex circuit, it is often advantageous to repeat portions of the circuit. If you are familiar with block copy functions in word processing, you will recognize Step as the two dimensional version of this. It duplicates the contents of the region defined by a Box. It copies these contents any specified number of times in either a vertical or a horizontal direction, or both.

Info

This provides information on any clicked component. If, for example, the component is a transistor, selecting the *Info* mode displays the library parameters of that transistor.

Windows Menu

You can exert some forms of window control without using this menu. In particular, you can move a window, close (erase) a window, scroll a window, or resize a window.

If you click and drag your mouse in the title bar of the window, you move the window. If you click the mouse in the close box (the square box located to the left of the title bar), you remove the front window. If you click the right button and drag the drawing with your mouse, you scroll the drawing. You can resize a window using the icon containing several sizes of rectangles located to the far right of the title bar. A click-and-drag operation originating on this icon will cause the window outline to track the mouse movement.

KEYBOARD Moving windows is accomplished by pressing Alt+F7 and then using the cursor keys to move the window. Terminate by pressing ⏎. You can remove the front window by pressing Esc. Scrolling is accomplished using the cursor keys provided the Scroll Lock key is on. Resizing is accomplished with Alt+F8 and then using the cursor keys. Terminate by pressing ⏎.

We now return to the pull-down menus at the top of the screen. The Windows menu controls the three main windows used in the program. These are the Mode window (at the left), the Component window (at the right), and the Circuit window. The pull-down menu contains a check if the corresponding window is visible. The menu is used to select which window to place at the front.

View Menu

This pull-down menu provides nine functions:

1:Restore windows: If you have altered the window size or placement, this command restores the windows to their standard size and location.

2:Scale 1:1: Displays the front circuit at normal scale.

3:Scale 2:1: Displays the front circuit at half size.

4:Scale 4:1: Displays the front circuit at one quarter size. Most of your circuits will probably fit in one screen. However, the full drawing area is made up of a 2 x 2 matrix of four standard-size drawing screens. Thus, you can scroll from one screen to another. In the Scale 4:1 mode, you can view the entire drawing area. This is important for more complex circuits.

5:Show comp text: Displays component parameter text when checked.

6:Show grid text: Displays grid text when checked.

7:Show node numbers: Displays node numbers when checked.

8:Search: Allows searching for a component device, component parameter, grid text, or node number. Not available in Student Edition.

9:Repeat search: Not available in Student Edition.

Editors Menu

This menu accesses the Library editors. This is very important as you use active devices in your circuits. There are two main libraries: devices and components. The devices library contains op-amps; diodes; bipolar transistors; MOSFETs; JFETs; and pulse, sinusoidal, and polynomial sources. The components editor controls the name, shape, parameter text location, electrical definition, and pin assignments for each component in the library. You will probably not be making any modifications to the components library, but you may wish to change some parameters in the device library. We discuss the most important of these below.

Operational amplifiers: The model of the operational amplifier contains 12 parameters, as follows:

Name	Parameter	Units
RIN	Input resistance	Ohms
AO	Open-loop gain	—
ROUT	Output resistance	Ohms
VOFF	Offset voltage	Volts
TC	Temperature coefficient for VOFF	V/°C
F1	First pole location	Hz
F2	Second pole location	Hz
SR	Slew rate	V/sec
IOFF	Input offset current	Amp
IBIAS	Input bias current	Amp
DI	Current doubling interval	°C
VMAX	Maximum output voltage	Volts

Bipolar transistors: Two bipolar transistor models are available: Gummel-Poon and Ebers-Moll. The Ebers-Moll model is mathematically the same as the Gummel-Poon model with certain second-order effects ignored. The Gummel-Poon model adds base width modulation (Early effect), low-current drop in beta, high-level injection, current-dependent base resistance, and transit time variation with collector current. Because the Ebers-Moll model is simpler, your simulation will run about 20% faster than it will with the more complex Gummel-Poon model.

The models contain 39 parameters that affect the simulation. These are listed below, along with the units and the default values. These can be modified from the Device library menu.

Name	Parameter	Units	Default
BF	Maximum Forward β	—	100.0
BR	Reverse β	—	1.0
XTB	Temperature coefficient for β	—	0.0
IS	Saturation current	Amps	1E-16
EG	Energy gap	eV	1.11
CJC	BC zero-bias depletion capacitance	Farads	0.00
CJE	BE zero-bias depletion capacitance	Farads	0.00
RB	Zero-bias base resistance	Ohms	0.00
RC	Collector resistance	Ohms	0.00
VAF	Forward Early voltage	Volts	1E30
TF	Forward transit time	Seconds	0.00
TR	Reverse transit time	Seconds	0.00
MJC	BC grading coefficient	—	0.33
VJC	BC built-in voltage	Volts	0.75
MJE	BE grading coefficient	—	0.33
VJE	BE built-in voltage	Volts	0.75
CJS	Collector-substrate zero-bias capacitance	Farads	0.00
VAR	Reverse Early voltage	Volts	1E30
NF	Forward emission coefficient	—	1.00
NR	Reverse emission coefficient	—	1.00
ISE	BE saturation current	Amps	0.00
ISC	BC saturation current	Amps	0.00
IKF	Corner for Forward β high-current roll-off	Amps	1E30
IKR	Corner for Reverse β high-current roll-off	Amps	1E30
NE	BE leakage emission coefficient	—	1.5
NC	BC leakage emission coefficient	—	2.0
RE	Emitter resistance	Ohms	0.0
IRB	Current where base resistance falls by half	Amps	1E30
RBM	Minimum base resistance at high currents	Ohms	RB
VTF	VBC dependence of TF	Volts	1E30
ITF	IC dependence of TF	Amps	0.00

Name	Parameter	Units	Default
XTF	Coefficient dependence of TF	—	0.00
XCJC	Fraction of BC depletion capacitance to internal base node	—	1.00
VJS	Substrate-junction built-in potential	Volts	0.75
MJS	Substrate-junction grading coefficient	—	0.00
XTI	Saturation current temperature exponent	—	3.00
KF	Flicker-noise coefficient	—	0.00
AF	Flicker-noise exponent		1.00
FC	Forward-bias depletion coefficient		0.50

Pulse sources: The pulse source produces a waveform as shown in Figure 23.

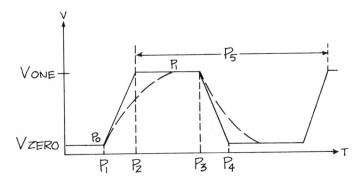

Figure 23

This is specified by seven parameters, as follows:

VZERO is the zero level or the initial value of the waveform in volts. Default value of 0.0.

VONE is the one level or the pulse value in volts. This parameter models the high level of the waveform and the default value of 5.0 volts.

P1 is the delay time in seconds. This parameter models the time delay from time equal zero to the leading edge of the waveform. Default value is 1.0E-7 seconds, or 0.1 microsec.

P2 is the time delay to the one level, that is, the time at which the high value is reached. The rise time is the difference between P2 and P1. Default value for P2 is 1.1E-7, yielding a rise time of 10 nsec.

P3 is the time delay to the start of the trailing edge. Its default value is 5.0E-7.

P4 is the time at which the low value is reached. The fall time is the difference between P4 and P3. Default value for P4 is 5.1E-7, yielding a decay time of 10 nsec.

P5 is the period of the waveform. Default value is 1.0E-6, or 1 microsec.

Sinusoidal sources: This provides for a sinusoidal source, with either constant amplitude or exponential decaying amplitude. The equation is given by (for TAU not equal to zero)

$$V(t) = A*e^{(-t/TAU)}*sin(2*(F + FS*t)*\pi*t + PH) + DC$$

where t is the time modulo the period; that is, t is the time from the start of the period. The parameters are listed below.

Name	Parameter	Units	Default
F	Frequency	Hz	1E6
A	Amplitude	Volts	1.0
DC	DC level	Volts	0.0
PH	Phase shift	Radians	0.0
TAU	Exponential time constant	Seconds	0.00
FS	Frequency shift term	Hz/sec	0.00

RS is the source resistance in ohms. You should never enter zero for this parameter. If you do not want the source to load the input signal, a small value of RS should be used.

RP is the repetition period in seconds. This parameter is used only when the exponent is nonzero and determines the time between successive applications of the exponential factor.

When the exponential time constant and frequency shift are set to zero (as in the default case), the sinusoid has constant amplitude. The equation is given by

$$V(t) = A*\sin(2*F*\pi*t + PH) + DC$$

Polynomial sources: The polynomial source computes the output variable as a polynomial function of the input variable. The general form of the polynomial equation is

$$Y(X) = A + B*X^C + D*X^E + F*X^G$$

where X is the input variable, Y is the output variable (current or voltage), and A through G are any real numbers. The type parameter is set to 0, 1, 2, or 3, as follows:

Type Parameter	Source Configuration
0	Voltage-controlled current source
1	Current-controlled current source
2	Voltage-controlled voltage source
3	Current-controlled voltage source

In the case of a current-controlled source, a 0.001 resistor is connected across the controlling nodes.

File Menu

This menu provides nine possible actions:

1:Create new circuit: Opens a new circuit window.

2:Load circuit: Loads a circuit file from disk. When selected, you are presented with a file dialog box that allows you to load, erase, or cancel the instruction. You scroll through available files from this box (see Figure 24). You can use this option to erase a file from the disk.

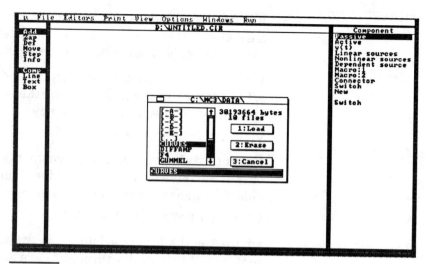

Figure 24

3:Unload circuit: Removes the front circuit window from the working memory (not from the disk). Once you use this option, you will need to select *1:Create new circuit* before proceeding to draw a new circuit.

4:Save circuit: Saves the front circuit to a disk file using the circuit name shown in the title bar of the window.

5:Save circuit as: Saves the front circuit to a disk file after requesting a new file name.

6:Merge circuit: Merges a circuit from disk with the front window circuit at the cursor position.

7:Load library: Loads a new device library from disk.

8:Save library as: Saves the library in memory to disk under a new name.

9:Change data path: Alters the data path for library, circuit, and output files.

Print Menu

This menu provides five print options:

1: Print Circuit: Allows you to print only the circuit.

2: Plot Circuit: Not available on the Student Edition.

3: Print front window: Prints a graphics image of the front window only. Alt+F3 prints the visible contents of this window at a smaller scale.

4: Print entire screen: Prints a graphics image of everything you see on the screen. Alt+F1 uses a large scale, and Alt+F2 uses a smaller scale.

5: Print netlist: Creates a netlist showing the contents of the current circuit. It sends the netlist to the destination specified in the *Text setup* function located on the Options menu.

Options Menu

This menu presents seven choices:

1:Palette: Not available on the Student Edition.

2:Preferences: Provides nine selections:

1. *Mouse ratio*: High or low ratio of mouse movement to arrow movement.

2. *Sound*: Enables or disables sound in warnings and error messages.

3. *File warning*: Enables or disables use of warning message when circuit has been changed.

4. *File dialog box*: Enables or disables file dialog box.

5. *Quit warning*: Causes system to ask for exit verification before quitting.

6. *Scale color*: Not available on the Student Edition.

7. *Floating nodes*: Checks circuit for floating nodes (connected to only one component) prior to analyses.

8. *Pivot solver*: Usually disabled, this routine is needed in some AC analysis.

9. *Node snap*: When enabled, components are adjusted so that pins coincide with nearest circuit component that is within two grids.

3:Global settings: Allows you to set eight values for use in analysis:

1. *DC Relative error*: Used to determine when the iterative process, used to solve the nonlinear DC equations, has converged. Typical values are between 1E-3 and 1E-6.

2. *Number of iterations*: Tells the system how many times to iterate a solution before regarding it as unsolvable and aborting the analysis.

3. *Relaxation factor*: Used in the operating point calculation. Typical values are between 1.1 and 2.0. If a particular circuit does not converge, make the value 1.2 or 1.1. This applies only to circuits with FET devices.

4. *Maximum voltage*: Used to aid DC convergence. It should be set between 1E3 and 1E6.

5. *Minimum conductance*: The minimum conductance of any component. Typical values are from 1E-9 to 1E-15.

6. *Switch on-resistance*: The resistance value that a switch will assume in the on state.

7. *Switch off-resistance*: The resistance value that a switch will assume in the off state.

8. *Std. dev. in tolerance*: Used for Monte Carlo analysis. Not available in the Student Edition.

4:Models: You can select models for bipolar transistors, MOSFET devices, and op-amps with this menu.

Two bipolar transistor models are available: the Ebers-Moll and the Gummel-Poon. The Ebers-Moll model is simpler than the Gummel-Poon since certain second-order effects are ignored. The Gummel-Poon model includes base width modulation, low-current drop in beta, high-level injection, current-dependent base resistance, and transit time variation with collector current. Because of this additional complexity, simulation time is about 20% greater with the Gummel-Poon. For most applications, the Ebers-Moll model will provide sufficiently accurate results.

Three MOS models are available: Level 0, Level 1, and Level 3. Level 0 is the simplest and will result in the fastest simulations. It has constant capacitance. If you have a long channel device, you will need to use Level 1. Level 3 provides the most accuracy but requires that you know many of the detailed parameters of the device. Levels 1 and 3 are the standard models found in SPICE programs.

Although the Model options menu contains two MESFET models, MESFETs are not available in the Student Edition.

Two op-amp models are available: Level 1 and Level 2. Level 1 is the ideal operational amplifier, and Level 2 includes 12 of the most common op-amp parameters. We discussed these parameters in the section "Editors Menu."

5:Plotter setup: Not used in the Student Edition.

6:Text setup: Text or numeric output generated by the program, such as netlists and analysis results, may be sent to a disk file, a window on the screen, or the printer. Printer output may be directed to one of several ports, in either dot matrix format or HP laser printer format.

7:Graphic setup: Printer output of the various graphics may be sent to a disk file or to the printer. Printer output may be directed to one of several ports, in either dot matrix format or HP laser printer format.

We suggest that you stop at this point and use options 6 and 7 to set up the program for the printer you are using (Epson or Laserjet). If you fail to do this, you may run into trouble when you try to print a circuit or simulation later.

μ Menu

This menu provides five selections:

1:Micro-Cap III: A title screen that shows the date, time, authors, and software revision letter.

2:Calculator: Provides an expression-type calculator. To use it, type in an expression such as

2.2*SIN(36.3) / EXP(–1.4)

Then press ↵ and the answer (–8.79028) will be returned. The operators supported by the calculator are shown in the section "v(t) Sources."

3:Help: A general help window. To get specific help related to the window you are using at any particular time, press F1.

4:DOS shell: Used to temporarily suspend MC3S and return to DOS, where you can run other programs. It is available only on 386 versions of MICRO-CAP III.

5:Quit: Used to exit the program. You can also press F3 to exit.

Conclusion: You should now be in a position to draw any circuit on the screen, save it, retrieve it, modify it, annotate it, and do all operations prior to the analysis.

In the next four tutorials, we deal with the four types of analyses you can perform on your circuit: transient, AC (bode plot), DC, and Fourier.

Problems

1. You have the following sample circuits on your disk: STD, SAMPLE, CURVES, PRLC, DIFFAMP, GUMMEL, L1. Retrieve each of these, print the circuit on your printer, and then redraw it yourself.

2. You wish to create the AM waveform,

   $\sin(2\pi 20000t)\sin(2\pi 10^6 t)$

Design a circuit that will produce this waveform as an output. You may use any type of sources and components.

3. Draw the circuit of Figure 47. Then store the circuit in an appropriately labeled file. You will need to retrieve it when you solve Problem 2 in Tutorial 2.

4. Draw the circuit of Figure 48. Then store the circuit in an appropriately labeled file. You will need to retrieve it when you solve Problem 3 in Tutorial 2.

5. Draw the circuit of Figure 49. Then store the circuit in an appropriately labeled file. You will need to retrieve it when you solve Problem 4 in Tutorial 2.

6. Draw the circuit of Figure 50. Then store the circuit in an appropriately labeled file. You will need to retrieve it when you solve Problem 5 in Tutorial 2.

7. Draw the circuit of Figure 71. Then store the circuit in an appropriately labeled file. You will need to retrieve it when you solve Problem 3 in Tutorial 3.

8. Draw the circuit of Figure 73. Then store the circuit in an appropriately labeled file. You will need to retrieve it when you solve Problem 5 in Tutorial 3.

9. Draw the circuit of Figure 75. Then store the circuit in an appropriately labeled file. You will need to retrieve it when you solve Problem 9 in Tutorial 3.

10. Draw the circuit of Figure 88. Then store the circuit in an appropriately labeled file. You will need to retrieve it when you solve Problem 3 in Tutorial 4.

11. Draw the circuit of Figure 89. Then store the circuit in an appropriately labeled file. You will need to retrieve it when you solve Problem 5 in Tutorial 4.

2

Transient Analysis

Introduction

If you have gone through Tutorial 1, you can now construct any circuit and draw it on the monitor. Although we have made the input process as simple as possible, drawing a circuit on the computer screen will probably take longer than on a piece of paper. You will realize the advantages of MICRO-CAP III, however, after you've drawn your circuit and begun to analyze the network you have created. This tutorial and the next three explore the various forms of analysis possible with the Student Edition of MICRO-CAP III.

Transient analysis is used to plot time waveforms at various points in the circuit. It involves generating a new set of equations dynamically for each time point, solving these equations, printing and graphing the solutions, and setting up a new set of equations whose content depends on the prior solution. Transient analysis evolves from the state space approach, which you may have been exposed to in your systems courses.

You begin by drawing a circuit, either by entering each component as described in Tutorial 1 or by retrieving a network from the data file. Although you may use any circuit you have created, we illustrate this for the differential amplifier circuit (one of the files on the disk we supplied). Recall this file from memory by using the File pull-down menu and selecting DIFFAMP. The screen should look like Figure 25.

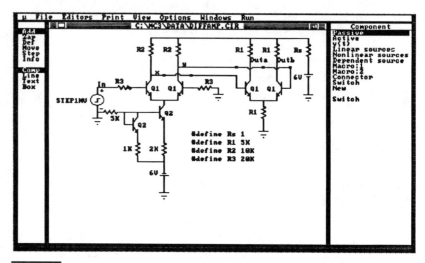

Figure 25

Once you have the network on the screen, you initiate the analysis by pulling down the Run menu and selecting transient analysis.

KEYBOARD Type **R** for Run followed by **1** (the number for transient).

Running the transient analysis produces the Transient analysis limit window. We discuss this in the next section.

To prepare you for the remainder of this tutorial, we now present a table of contents.

Analysis Limits

Transient Analysis Menu Selections

    Run

We begin with a discussion of the Analysis limits window, which you see as soon as you activate the transient analysis simulation. We then discuss each of the menu selections in the transient analysis.

Analysis Limits

After activating the transient analysis, you are presented with a window showing eight limits used in the analysis. When you store a circuit, the limits are stored with it. If you create a new circuit, default limits are used. The limits stored in the differential amplifier circuit are shown in Figure 26.

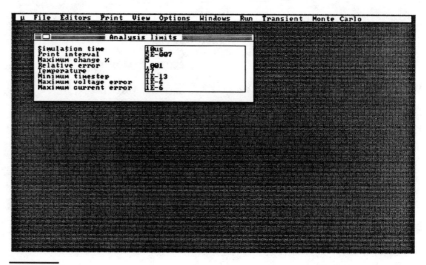

Figure 26

We take the time now to describe each of these analysis limits. However, if you are impatient to see an analysis run, you can simply pull down the Transient menu and select *Run* (or press F2).

Simulation time: This is the time over which the simulation is performed. Three parameters can be specified: the starting time (Tmin), the ending time (Tmax), and the maximum time increment used in the iteration analysis (Maxtimestep). If the maximum time increment is too large, the resolution suffers, whereas if it is too small, the program takes a long time to run. The program adaptively chooses step size based on truncation error and rate of change of the waveform. It will never use steps larger than the specified Maxtimestep.

The format for simulation time is

Tmax [,Tmin] [,Maxtimestep]

The two terms in brackets are optional, and if left off, the default values are used.

The default value for Tmin is 0, and the default value of Max-timestep is 1% of the total simulation time [.01*(Tmax–Tmin)].

Print interval: This parameter is used only if you are printing a table of outputs in numeric form (we see how to do this when we discuss options). The program prints the parameter values at time samples separated by the print interval. This parameter has no effect on the various plots.

Maximum change %: We previously indicated how the simulation selects sample points. However, if the waveform values change by more than this specified percent of full scale, the system reduces the time step. The maximum change % therefore affects the accuracy of the final simulation. Values generally run between 1% and 5%, with the smaller values used for circuits that produce waveforms with steep slopes.

Relative error: During the analysis of a nonlinear circuit with the *Iterate* option enabled (see the section "Analysis Options"), the system finds the solution at each time point using Newton-Raphson iteration. Convergence is achieved only when successive solutions at each time point agree to within a tolerance equal to the specified relative error. A typical value is 1E-3.

Temperature: One or more temperature values can be specified for the analysis. The format is

Final [,Initial[,Step]]

The default value for Initial is Final, and the default value for Step is the difference between Final and Initial. Therefore, if only one value is specified, the simulation will run at that temperature. If two values are specified, the simulation will run at these two values, the Initial and Final temperatures. For example, 27, 25 would run the simulation at both 25°C and 27°C (27°C is room temperature). An entry 35,20,5 would run the simulation at 20°, 25°, 30°, and 35°.

Minimum timestep: This is the smallest time step allowed during the analysis. Typical values are 1E-15 to 1E-13, but this depends on the time constants of the circuit. If you choose a smaller value than needed for the desired output resolution, the simulation will take longer than necessary to execute.

Maximum voltage error: This is the absolute error tolerance for voltage.

Maximum current error: The absolute error tolerance for current.

If you wish to change any of the analysis limits, click the mouse on that limit, and then type in the new limit.

KEYBOARD To change an analysis limit using the keyboard, position the cursor using the cursor arrow keys, and then edit the entry.

Once you are satisfied with the analysis limits, you are ready to specify what you want plotted and then to run the analysis. Since we are using the example of the differential amplifier, let us first run the simulation using the variables that have already been selected for plotting. Run the analysis by pulling down the Transient menu and clicking *Run*. You can also press F2. The transient analysis is then produced, as shown in Figure 27.

We have plotted four waveforms on these two graphs. (We learn how to specify what is plotted in the section "Monitor.") The time scales are common, but each of the four waveforms has a different vertical axis. We are plotting three voltages (the input, the voltage between x and y, and the voltage between Out a and Out b). We are also plotting a power waveform showing the power supplied by the 6V battery.

We now see how to select and specify the waveforms for plotting.

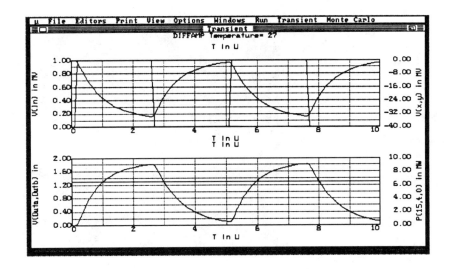

Figure 27

Transient Analysis Menu Selections

Once you select Transient analysis from the Run menu, in addition to presenting the analysis limits, two pull-down menus are added to the top of the screen: the Transient and the Monte Carlo analysis. We describe the entries in the Transient menu in this section. Monte Carlo analysis is not supported in the Student Edition.

Run

Selecting this menu entry runs the simulation.

Limits

Note that this menu entry is checked. It produces the analysis limits if you wish to change them later.

Options

Clicking this menu entry presents a window, as shown in Figure 28.

The window is divided into four areas: *Run options*, *Initial values*, *Analysis options*, and *Other options*. Options may be selected by clicking the mouse. You can leave the Options window either by initiating the run (pull down Transient

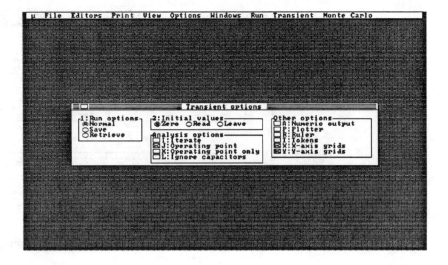

Figure 28

menu and select *Run*) or by erasing the window by clicking
on the box to the left of the title.

KEYBOARD   Select options using either the Tab key or the number or letter adjacent to the desired item or group. To select an unnumbered or unlettered option in a group, use the up and down
arrow keys to select the option and the space bar to activate
the selection. Use the Esc key to exit the Options window, or
press F2 to start a simulation run.

Run Options

Normal produces a normal run. *Save* saves the entire simulation for later retrieval. *Retrieve* retrieves a saved simulation
for review.

Initial Values

Zero sets the initial capacitor voltages and inductor currents
to zero prior to each temperature or step run. *Read* reads the
initial voltages and currents from a previously saved disk file
prior to each temperature or step run. *Leave* does not change
the initial voltages and currents prior to each run; they

remain unchanged from either the prior run, the last file read, or the last edit.

Analysis Options

Iterate forces the simulator to iterate at each time point. Most nonlinear circuits require iteration to produce accurate results, but linear circuits normally do not require iteration. Since iteration increases the run time, we recommend first trying the simulation with the iterate disabled.

Operating point forces the simulator to calculate the DC operating point prior to each run due to the t = 0 values of the sources. It is needed for most nonlinear circuits.

Operating point only forces the simulator to calculate the DC operating point and terminate the run. This can be used to find quiescent values in a circuit. The DC values are available from the initial values editor following a simulation run.

Ignore capacitors causes the transient analysis to ignore capacitor currents. This is useful if you are interested in DC values rather than transient spikes caused by step voltages across capacitors.

Other Options

Some of these are quite useful to distinguish plots when using a noncolor printer.

Numeric output: Causes selected waveforms to be sent to a disk file, the screen, or the printer, according to the selection under Text setup (see the section "Options Menu"). The destination file is named CIRCUITNAME.TNO, where CIRCUITNAME is the name of your circuit. The selected waveforms are specified in the Monitor menu, which we cover in the section "Monitor."

Plotter: Not available in the Student Edition.

Ruler: Selects ruler marks instead of full-screen graph divider lines.

Tokens: Adds a token to the second waveform of each graph. The token also appears on the corresponding axis scale.

X-axis grids: Draws horizontal grids or ruler marks.

Y-axis grids: Draws vertical grids or ruler marks.

Scope

This option allows you to examine the simulation curves in more detail. After you run the simulation, select the *Scope* option from the Transient menu, or press F8. The resulting scope display, for the DIFFAMP example, looks like Figure 29.

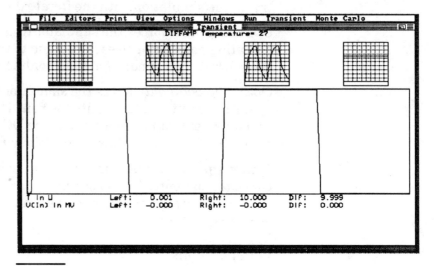

Figure 29

The top portion of the screen shows a plot icon of each of the four waveforms. The bottom portion shows a blowup of one or more of these waveforms. Initially, the first waveform is displayed, and the lower section of that icon is highlighted. You can select or add other waveforms by clicking on the lower section of the appropriate icon. If we click the lower section of the second curve, it appears in the lower graph, as shown in Figure 30.

Both the first and second curve appear. If we do not want the first curve, we click on the lower section of that icon, and only the second curve is plotted on the lower portion of the screen.

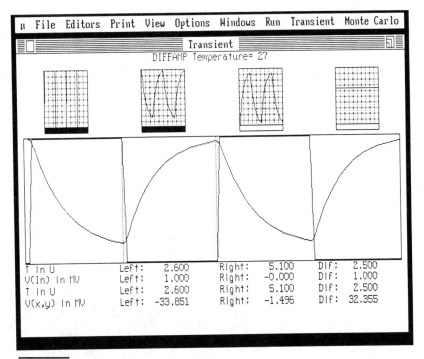

Figure 30

The lower display has two vertical cursors. These can be used as markers to read values from the curve. In Figure 30, we have set the left cursor (the dotted vertical line) at the minimum of the V(x,y) waveform, and the right cursor is at the maximum. The value of these cursors is that the X and Y numerical values are printed below the graph. For example, the minimum of V(x,y) occurs at 2.6 usec and a value of –33.851 mV. The maximum occurs at a time of 5.1 usec and a value of –1.496. The printout also presents the difference between these two values. The cursor positions are set by clicking the left mouse button at the desired location of the left cursor and the right mouse button at the right cursor.

KEYBOARD   You can use the < and > keys to move the selected cursor left and right. Use the Tab key to toggle between the left and right cursor.

You can also zoom in on portions of the curve. You define a box on the icon, and that box fills the lower screen. You position the mouse arrow at the upper left of the desired box, and click the left button. Then move to the lower right, and press the right button. Alternatively, you can use the + and - keys to resize the box and the Shift+cursor keys to relocate the box. Try this to see how it works.

The portion of the curve you wish to view may be outside the portion you have chosen to plot. You can use the cursor keys to scroll the plot.

Stepping

Component parameters may be stepped from one value to another, producing multiple runs with multiple waveforms or plots. You activate this feature by selecting *Stepping* from the Transient menu and then entering instructions in the dialog box, which is shown in Figure 31.

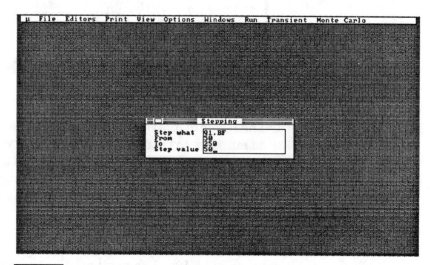

Figure 31

In some of the figures in this part of the manual, we have erased all but the front window. You will probably not be doing this, so the other active windows will show on your screen. Figure 32 shows this for the Stepping analysis window. You can always erase the front (active) window by clicking on the box to the left of the window title.

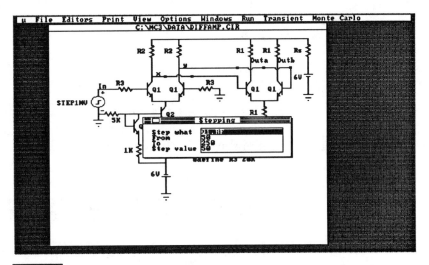

Figure 32

Step What

This is where you specify which parameter you wish to change. For elements, you would enter the name of the element. For example, if you named one or more resistors R1, you would enter **R1**. For devices, you enter the device name, a period, and the parameter you wish to vary. For example, if you used a 2N2222A transistor and wish to vary forward beta, you would type **2N2222A.BF**.

Limits

From specifies the starting value of the parameter. *To* specifies the ending value of the parameter. *Step value* specifies the amount to step the parameter by. As an example, suppose we use our DIFFAMP example circuit and vary transistor Q1 beta from 50 to 250 in steps of 50. That is, we require five runs with betas of 50, 100, 150, 200, and 250. We would select the *Transient* option and then select the *Stepping* menu entry. We type in these values:

Step what	Q1.BF
From	50
To	250
Step value	50

Then activate the run either by pulling down the Transient menu and selecting *Run* or by pressing F2. The resulting simulation is shown in Figure 33.

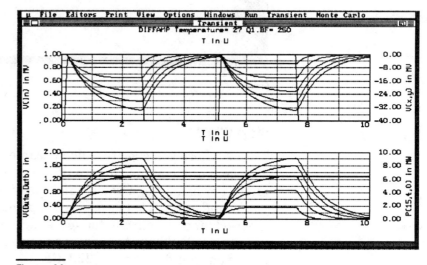

Figure 33

Palette

This is not supported in the Student Edition.

Monitor

This is one of the most important menu selections. It specifies which wavforms are to be plotted and the axis limits for each of these. When you select this menu item, a Monitor window appears. Figure 34 shows this window for the differential amplifier example.

The left portion of the window has four columns of boxes that can be checked:

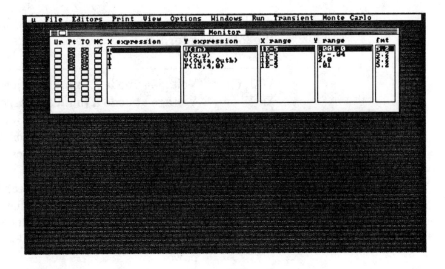

Figure 34

Ur: A check mark in this column tells the system to save the waveform in a user file source data file. We describe these in Tutorial 5. The file consists of a number of equidistant samples of the Y expression waveform specified in the row containing the check mark. Only one row can have a Ur check per run.

Pt: A mark in this column tells the system to plot the specified waveform on the transient graphs. Although the window contains ten rows, a maximum of eight waveforms can be simultaneously plotted (four per graph).

TO: A mark in this column tells the system to print the numeric value of the waveform to a screen window, a disk file, or the printer.

MC: This is for Monte Carlo; it is not supported in the Student Edition.

We then have five areas of this window to specify the various waveforms. The X expression and the Y expression define the expressions to be evaluated and plotted. Typically X would be the time variable, and you would simply enter **T**. Y could be a voltage, current, power, or energy and is specified in any of the following forms:

V(X[,Y])	Voltage at node X [minus voltage at node Y]
X[–Y]	Voltage at node X [minus voltage at node Y]
V(@Name)	Voltage across the resistor or inductor called Name
I(@Name)	Current through the resistor or inductor called Name
I(X,Y)	Current through the resistor or inductor whose nodes are X and Y
P(X,Y,Z)	Power flowing into a circuit section. X and Y define the nodes between which must exist a resistor or inductor that measures the current. The current is multiplied by the voltage drop between the nodes Y and Z to get power.
E(X,Y,Z)	Energy flowing into a circuit section. Same format as power.

The unit of voltage is volts, the unit of current is amps, the unit of power is watts, and the unit of energy is joules.

We then need to enter the X and Y range. The format for this is

Xmax [,Xmin] and Ymax [,Ymin]

If you do not specify the minimum, the default value is zero.

The final column is fmt, which specifies the numeric output precision. You enter this in the form

L.R

L specifies the number of digits to the left of the decimal point, and R specifies the number of digits to the right. In the differential amplifier example, we use 5.2, which will use numbers with five digits to the left and two to the right of the decimal point.

Initial Values

We mentioned in the introduction that the simulation uses a state-space approach toward stepping through the circuit solution. The state values are the node voltages and inductor currents. You can use the Initial values editor to review and edit these values. When you select this menu entry, an Initial values window appears on the screen. A sample is shown for the DIFFAMP example in Figure 35.

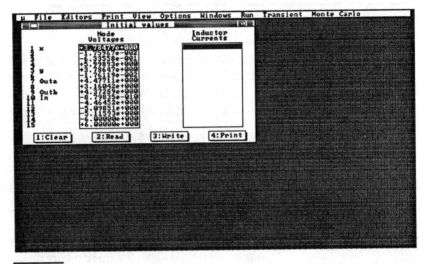

Figure 35

These initial values are those in memory for this circuit. They are the ending values of the previous analysis. If you have not yet run an analysis of the circuit, they will usually be zero. In the section "Options," we saw that there are three choices: the initial values are those calculated; all zero; and values read in from a disk file. These initial values are used prior to the calculation of the operating point.

You can edit the initial values by selecting a value with the mouse and typing the new value.

Use the Tab key to change fields horizontally and the cursor keys to change vertically.

At the bottom of the window are four commands: *Clear* sets all values to zero; *Read* reads the values from a disk file; *Write* writes the displayed values to a disk file; *Print* prints the values to the standard Text output location.

Saving and Printing the Simulation Results

You can save the simulation results for analysis at a later time. One type of analysis you may wish to perform is Fourier analysis. We defer discussion of saving waveforms until Tutorial 5, where we cover Fourier analysis in detail.

Printing the analysis output is simple using the Print pull-down menu. We discussed this menu in the section "Print Menu" in Tutorial 1. Your options are either to print the front window (which would print just the graphics image) or to print the entire screen (which includes everything you see, including the menu bar at the top of the screen). You can also print either of these in a smaller scale using Alt+F3 (for front window) or Alt+F2 (for entire screen).

If your printer does strange things (for example, feeds many sheets of paper, each with only a few characters on it), you might not have set the program for the proper printer. Refer to the section "Options Menu" in Tutorial 1 for the *Graphic setup* options.

Examples

1. You are given the emitter follower amplifier of Figure 36.

   a. Determine the values of V_{CEQ} and I_{CQ}.

   b. Find the maximum symmetrical output voltage swing.

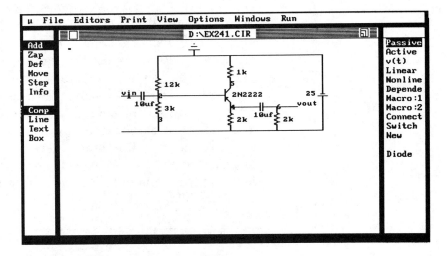

Figure 36

Solution: We begin by entering the circuit into the computer. This is done in a straightforward manner using the techniques of Tutorial 1. There are alternate ways to enter the component values; for example, the 10-microfarad capacitors can be entered as 10u, 10uf, 1E-5, or .00001. Depending on how you enter the circuit, your node numbers may differ from those shown.

(a) We can find the quiescent values of voltage and current by running a transient analysis without placing any source at the input. Since you will have to specify the variables you wish to plot (using the Monitor menu), you should get into the habit of using the View menu to select the mode where node numbers are displayed. Then print the circuit with these node numbers. This will make it much easier to specify the variables for display.

The collector-to-emitter voltage is that between nodes 5 and 4, so we specify this in the transient monitor. The collector current is the current through the 1 kΩ resistor between nodes 0 (ground) and 5, so we specify I(0,5). If there had been no inductor or resistor in this branch, we would have to add a small component to measure the current. In setting the limits, you can use trial and error, or you can refer to electronic theory. The magnitude of the voltage will certainly lie between 0 and the source voltage, so we can set that range to 0 to –25. The voltage across the collector resistance cannot exceed 25/3 volts (when the transistor is saturated and we have a voltage divider with the 2 kΩ resistor in the emitter leg). We therefore can set the current range from 0 mA to –10 mA. You can refine these ranges after the first run. The time range is not important since there is no transient to die out. We have chosen to leave it at 1 microsecond. The result of the transient analysis is shown in the upper plot of Figure 37.

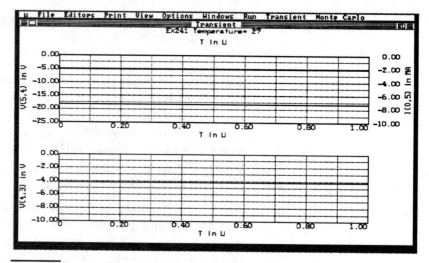

Figure 37

The quiescent collector-to-emitter voltage is approximately –18 volts, and the quiescent collector current is –2 mA. You can read these values more accurately using the scope display.

(b) Since this circuit is an emitter follower, the output voltage is approximately equal to the input. You can verify this by hooking any appropriate source to the input (for example, a sinusoid) and then displaying input and output waveforms. We have displayed the quiescent emitter voltage [V(4,3)] on the second graph. This voltage must remain negative so that the transistor conducts. Therefore, the maximum swing is approximately 67% of 3.7 V for a peak-to-peak voltage of about 4.9 volts. We can verify this by placing a sinusoid at the input and varying its amplitude. For example, we can select the 1VPP sinusoidal voltage from the menu of sinusoidal sources. This is a 2 kHz sinusoid with an amplitude of 0.5 volts. If we add this to the input nodes and select a stepping analysis, we can vary the amplitude of the source around the expected peak. In Figure 38, we have plotted the input voltage and output voltage where we step the sinusoidal amplitude from 2 V to 4 V in steps of 1 V. We do this by specifying 1VPP.A when asked, Step what. Since the sinusoid has a frequency of 2 kHz (that is, a period of 0.5 msec), we have modified the time range to varying from 0 msec to 1 msec. We do this on both the Monitor and the Limits menus. We have also inverted one of the axes so that the curves for input and output do not overlap. Once the amplitude exceeds approximately 2.2, the output is distorted by being clipped. This is shown in Figure 38. The output waveform saturates at 2.2 V in the positive direction,

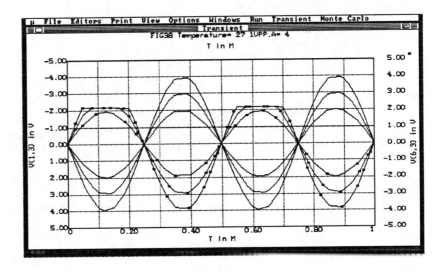

Figure 38

but is not distorted in the negative direction. If we were asked to design this circuit for maximum output swing, this result would indicate a poor design (that is, the Q-point is not in the middle of the load line).

2. For the common emitter amplifier shown in Figure 39, how much power is dissipated in the transistor?

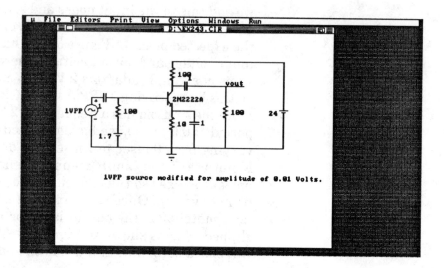

Figure 39

Solution: We enter the circuit component by component. Since we wanted a 0.02 V peak-to-peak 2 kHz sinusoid, we modified the 1VPP source in the sinusoidal device library by changing A from 0.5 V to 0.01 V. The lack of a modifier on the capacitor values indicates 1-farad capacitors. These are not practical in a real-life circuit, but the large value assures that they act as short circuits for the AC components of the signal.

The power dissipated in the collector emitter is approximately equal to the total power dissipated in the transistor. We can display this directly on the transient analysis plot by specifying P(6,4,5) on the transient monitor. We are therefore specifying the collector current as that from node 6 to node 4 (through the 100 Ω resistor). The voltage is the voltage between nodes 4 and 5, the collector and emitter. The resulting plot is shown in Figure 40.

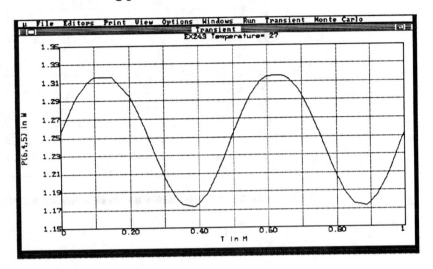

Figure 40

The average power is approximately 1.25 watts, and the peak power is almost 1.32 watts. The specifications for this transistor provide for a maximum power dissipation of 0.5 watts at 25°C ambient air temperature, and 1.8 watts at a maximum case temperature of 25°C. Therefore, with the choice of resistors in this circuit, we would have to provide a heat sink if we did not wish to exceed the specifications of this transistor.

3. For the CE amplifier shown in Figure 39, determine the variation of voltage amplification if β varies from 50 to 150.

   Solution: We have already entered the circuit, and there is no need to redraw it. We need simply select the stepping mode, and when asked, Step what, respond with 2N2222A.BF. We have chosen to step beta from 50 to 150 in steps of 50. The resulting input and output waveforms are shown in Figure 41. We can find the amplification by dividing the output amplitude by the input amplitude. (It

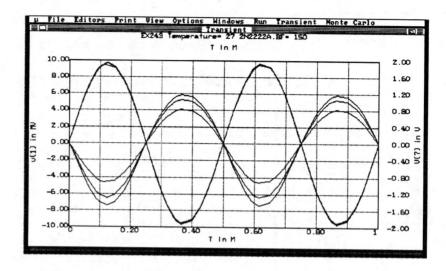

Figure 41

is easier to find amplification from the AC gain plot, but since this is covered in Tutorial 3, we use the more primitive time function method for this problem.)

The peak-to-peak input amplitude is approximately 0.019 V. The reason it is not exactly 0.02 V is the internal resistance of the source. The output peak-to-peak amplitudes for the three beta values are 1.72, 2.4, and 2.65 for amplitude gains of 90, 126, and 139, respectively.

4. Find the output waveform for the circuit shown in Figure 42, where the input is sinusoidal with an amplitude of 5 V and a frequency of 100 Hz.

    Solution: We enter the circuit component by component. You can either modify one of the sinusoidal sources or use the "Own" to define your own.

    In entering circuits with op-amps and other active components, it is important that you connect components and wires to the nodes of the active devices. We intentionally did not do so in Figure 42. Note that we have printed out node numbers and that node 2 should be the same as node 5. That the program printed two separate numbers indicates that at least one of our connections is not occur-

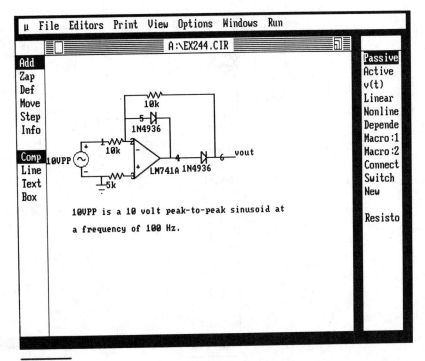

Figure 42

ring at the node. You can verify this by pulling down the Print menu and selecting *Netlist*. We can correct this either by redrawing the circuit or by adding additional lines to short the appropriate points. We illustrate this in Figure 43, where we have added an additional line and made the two nodes into one.

We now run the transient analysis. The result is the plot shown in Figure 44.

When dealing with op-amps and diodes, some precautions are necessary. Because of the extremely small input currents to op-amps, care must be taken in selecting components to connect to the input. For example, if you were to substitute 1N4001 diodes for the 1N4936 diodes of Figure 42, you would find that the outputs have discontinuities. This occurs because the library model for the 1N4001 contains 0 series resistance (an accepted approximation). This ideal approximation, coupled with

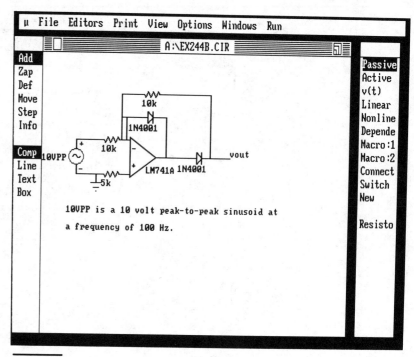

Figure 43

transients from the capacitors in the model, causes the discontinuities. You can correct this either by selecting *Ignore capacitors* in the Analysis options menu or by changing the model to contain a small resistance. The latter can be accomplished either by altering the library entry or by using a model statement of the form,

#Model 1N4001 D(RS=1)

where we have set the forward resistance to 1 Ω.

5. You wish the circuit of Figure 45 to operate as a full-wave rectifier with a gain of 1 for negative inputs and a gain of 3 for positive inputs. You solve the necessary equations, and obtain the resistor values shown in the circuit of Figure 45. Another student in your class has solved the same problem and obtained the same circuit, except that R_F, the feedback resistor for the second op-amp, is 350 kΩ. Your teacher has asked you to simulate the response to a sine wave and determine which solution is correct.

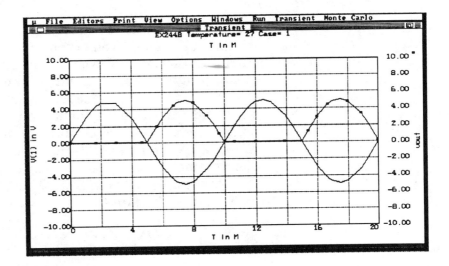

Figure 44

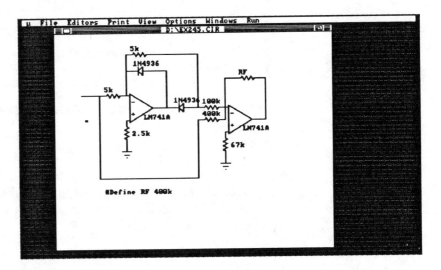

Figure 45

Solution: You enter the circuit component by component making sure that the connections to the op-amps are occurring at the nodes. You check this by numbering nodes and by printing a netlist. You then run the simulation. You can find the gain by feeding in any appropriate waveform. (An easier way to find gain is to use DC analysis, which we describe in Tutorial 4.) We have chosen a 5-V-amplitude 100-Hz sinusoid as in the previous example. You use the Stepping mode to step the feedback resistor between 350 kΩ and 400 kΩ. The result is the curves of Figure 46.

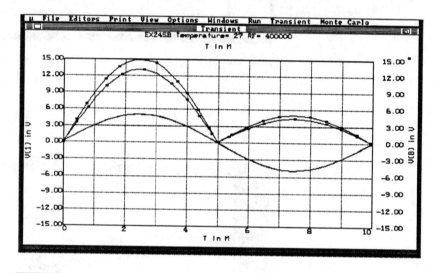

Figure 46

The upper output curve is generated for a resistor value of 400 kΩ and is clearly the one that represents the correct gains.

Problems

1. In Tutorial 1, you were asked to design a circuit to produce an AM waveform,

   $$\sin(2\pi \times 20000t)\sin(2\pi \times 10^6 t)$$

   Plot the time waveform for several cycles of the first sinusoid.

2. You have carefully designed the amplifier shown in Figure 47 to have a low temperature sensitivity. The circuit must operate between 20°C and 40°C. The response to a 1-mV pulse must not exceed 60 mV at any time since this is driving an A/D and you do not want to saturate. Use a simulation to find the best value for the collector resistor.

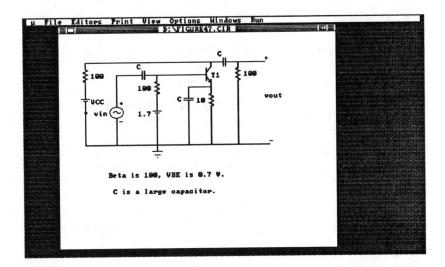

Figure 47

3. Find the range of quiescent collector currents for the amplifier shown in Figure 48. The transistor beta varies from 300 to 400, and the temperature varies from –50°C to +65°C.

   $$R_1 = R_2 = 4 \text{ k}\Omega$$

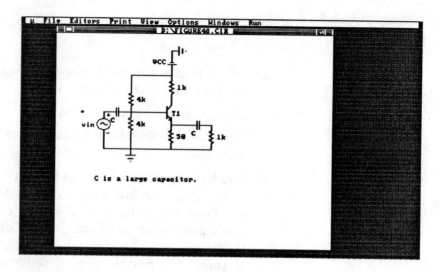

Figure 48

$R_E = 50\ \Omega$

$R_L = R_C = 1\ k\Omega$

Now let the input be a sinusoid of amplitude A. Find the maximum value of A before there is visible distortion in the output. Repeat for $R_E = 100\ \Omega$.

4. Find the output waveform of the circuit shown in Figure 49.

5. A 1-kHz square wave forms the input to the circuit of Figure 50. Plot the output waveform.

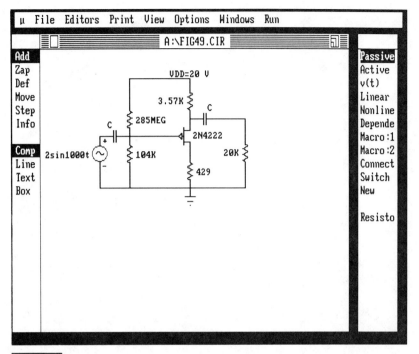

Figure 49

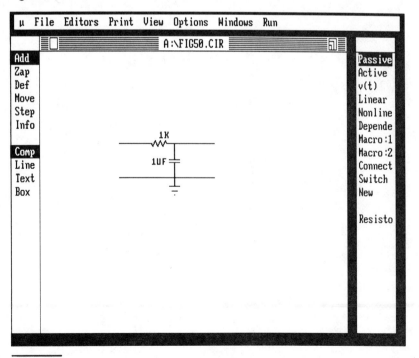

Figure 50

3

AC Analysis

Introduction

AC analysis evaluates the small-signal transfer characteristics of a circuit. Before running the analysis, the system replaces all voltage sources (time-varying sources and batteries) with 0.001 Ω resistors. If the circuit being analyzed contains any nonlinear elements, the program calculates the DC operating point to determine the small-signal characteristics of each nonlinear element. Each nonlinear element is replaced with linear elements representing these small-signal characteristics.

In performing the AC analysis, the system connects a 1-volt, variable-frequency, sine wave voltage source from ground to the specified input node. The system then varies the frequency of this source. For each AC frequency analysis sample, the system solves a series of equations to evaluate the output at the specified node. The output is then compared with the input. The system can plot any of the following parameters:

Gain in both linear and log format: The complex ratio of the output voltage to the input voltage.

Input impedance in ohms: The complex ratio of input voltage to input current.

Input admittance in mhos: The complex ratio of input current to input voltage.

Output impedance in ohms: The complex ratio of output voltage to output current.

Output admittance in mhos: The complex ratio of output current to output voltage.

Input and output noise in both linear and log format: This is a measure of the output voltage due solely to noise sources arising from resistors, bipolar transistors, MOSFETs, JFETs, and diodes. The linear plot is expressed in volts per square root of bandwidth. The log plot is in decibels, which is 20 times the log of the amplitude.

Nyquist diagrams: A plot of the imaginary part of the output versus the real part of the output.

Group delay: The slope of the phase characteristic curve, that is, the ratio of phase shift to frequency. The units of group delay are seconds.

Except for the noise and group delay, each of the above parameters is complex. It is represented by two plots: magnitude and phase.

You begin by drawing a circuit, either by entering each component as described in Tutorial 1 or by retrieving a network from the data file. Although you may use any circuit you have created, we illustrate this for the differential amplifier circuit (one of the files on the disk we supplied). Recall this file from memory by using the File pull-down menu and typing the name **DIFFAMP**, or by selecting it from the list that is presented. The screen should look like Figure 51.

Once you have the network on the screen, you initiate the analysis by pulling down the Run menu and selecting *AC* analysis.

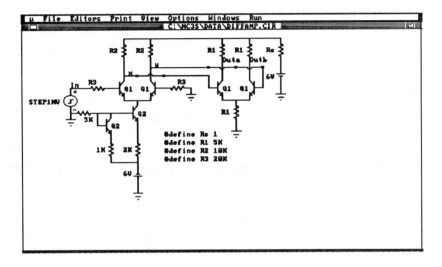

Figure 51

KEYBOARD Type **R** for Run followed by **2** (the number for AC).

Running the AC analysis produces the AC analysis limit window. We discuss this in the next section.

To prepare you for the remainder of this tutorial, we now present a table of contents.

Analysis Limits

AC Analysis Menu Selections

    Run

    Limits

    Options

        Run Options

        Monte Carlo

        Frequency Step

        Main Variable

        Vertical Scale

        Horizontal Scale

Analysis Limits

After activating the AC analysis, you are presented with a window showing 13 limits used in the analysis. When you store a circuit, the limits are stored with it. If you create a new circuit, default limits are used. The limits stored in the differential amplifier circuit are shown in Figure 52.

We take the time now to describe each of these analysis limits. However, if you are impatient to see an analysis run, you can simply pull down the AC menu and select *Run* (or press F2).

Frequency range: The format for this is

Fmax[,Fmin[,N]]

The analysis starts at Fmin and ends at Fmax. If the *Auto step* option is selected (see the section "Options"), the system automatically determines the frequency points at which to calculate a solution. If the *Fixed step* option is selected, the system calculates N solutions spaced equally over the frequency range, Fmax–Fmin (if the X scale is linear), or N solutions spaced equally over each decade (if the X scale is logarithmic).

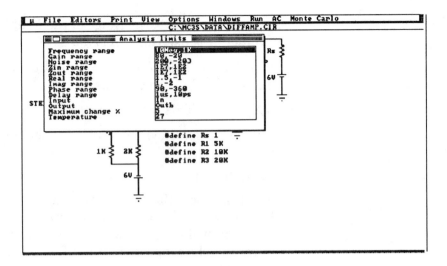

Figure 52

If you omit N, it defaults to 50 for linear X scales and to 10 for log X scales. If you omit Fmin, it defaults to zero for linear and to $Fmax/10^5$ for log X scales (if it defaulted to zero, the log would not be bounded).

Gain range: This sets the limits of the gain axis. The format is

Max[Min]

Default for the minimum is zero for linear and $Max/10^5$ for log scales.

Noise range: This sets the limits for the noise axis. Follows the same rules as Gain range.

Zin range: This range is used to scale the input impedance axis.

Zout range: This range is used to scale the output impedance axis.

Real range: This is the range of the real part of the variable. It is used only in Nyquist plots.

Imag range: This is the range of the imaginary part of the variable. It is used only in Nyquist plots.

Phase range: This sets the scale of the phase angle axis. Specify the phase in degrees.

Delay range: This sets the range for the scale of delays.

Input: This specifies the location of the input source. The format is

Plus[,Minus]

For Plus, insert the node name (number or text name) where the positive lead of the input source is to be connected. Minus is the name of the node where the negative lead of the input source is to be connected. The default if you omit the Minus is the zero or ground node.

Output: Follows the same rules as Input but is used to specify the location where the output is measured.

Maximum change %: During the analysis, the program automatically adjusts the frequency step up and down if the *Auto step* option is selected. If the plot values change by more than this percent of full scale, the system reduces the step size. Typical values needed to produce smooth curves range from 1% to 5%.

Temperature: The format for specifying temperature is

Final[,Initial[,Step]]

The simulation will be performed at temperatures between the Initial and the Final, spaced by Step. Thus, for example, an input of 35,20,5 would produce separate runs at 20, 25, 30, and 35 degrees centigrade. If Step is omitted, two runs are performed, one at the Initial and one at the Final. Thus, the default value of Step is Final minus Initial. If both the Step and Initial are omitted, one run is performed at the Final temperature. Thus, the default value of Initial is Final.

If you wish to change any of the analysis limits, click the mouse on that limit, and then type in the new limit.

To change an analysis limit using the keyboard, position the cursor using the cursor arrow keys, and then edit the entry.

Once you are satisfied with the analysis limits, you are ready to run the simulation. Let us first run the simulation using the variables already selected for plotting. Run the analysis by pulling down the AC menu and clicking *Run*. You can also press F2. The AC analysis is then produced, as shown in Figure 53.

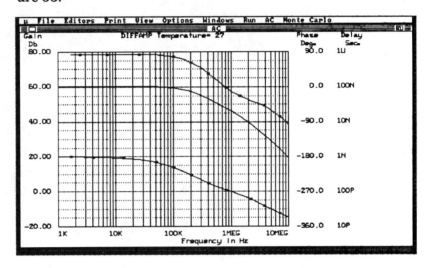

Figure 53

We now analyze the plotting parameters in more detail.

AC Analysis Menu Selections

Once you select *AC* analysis from the Run menu, in addition to presenting the analysis limits, two pull-down menus are added to the top of the screen. These are the AC and the Monte Carlo analysis. We describe the entries in the AC menu in this section. Monte Carlo analysis is not supported in the Student Edition.

Run

Selecting this menu entry runs the simulation.

Limits

This menu entry is checked. It produces the analysis limits if you wish to change them later.

Options

Clicking this menu entry presents a window, as shown in Figure 54.

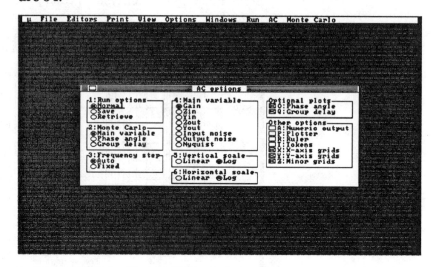

Figure 54

The window is divided into eight areas: *Run options*, *Monte Carlo*, *Frequency step*, *Main variable*, *Vertical scale*, *Horizontal scale*, *Optional plots*, and *Other options*. You can leave the AC options window either by initiating the run (pull down the AC menu and select *Run*) or by erasing the window by clicking on the box to the left of the title. You can also initiate the run by pressing F2.

KEYBOARD

Select options using either the Tab key or the number or letter adjacent to the desired item or group. To select an unnumbered or unlettered option in a group, use the up and down arrow keys to select the option and the space bar to activate the selection. Use the Esc key to exit the AC options window, or press F2 to start a simulation run.

Run Options

Normal produces a normal run. *Save* saves the entire simulation for later retrieval. *Retrieve* retrieves a saved simulation for review.

Monte Carlo

Monte Carlo analysis uses probability rules to control multiple runs. This option selects the parameter to be varied according to these probability rules. However, Monte Carlo analysis is not supported in the Student Edition of MICRO-CAP III.

Frequency Step

Auto enables the automatic frequency step control (see the section "Analysis Limits"). *Fixed* forces the frequency step to be a fixed value.

Main Variable

This selects which parameter is plotted in the main plot. You can select *Gain*, input impedance (*Zin*), input admittance (*Yin*), output impedance (*Zout*), output admittance (*Yout*), *Input noise*, *Output noise*, or a *Nyquist* diagram.

Vertical Scale

This selects whether the vertical scale should be linear or logarithmic (dB). Nyquist plots always use a linear scale.

Horizontal Scale

This selects whether the horizontal scale should be linear or logarithmic (dB). Nyquist plots always use a linear scale.

Optional Plots

You can select one or both of these optional plots. If you select *Phase angle*, the phase angle is plotted, and the corresponding vertical scale is placed at the right side of the graph. If you choose *Group delay*, this is also plotted with a scale placed at the right side of the graph, adjacent to the phase scale if you have also selected that option.

Other Options

Some of these are quite useful to distinguish plots when using a noncolor printer.

Numeric output: Causes selected waveforms to be sent to a disk file, the screen, or the printer, according to the selections under Text setup. If you have selected a screen output, following display of the curves, you are presented with a table of the type shown in Figure 55.

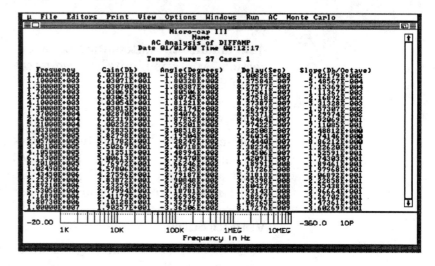

Figure 55

The destination file is named CIRCUITNAME.ANO. (ANO stands for AC Numeric Output.)

Plotter: Not available in the Student Edition.

Ruler: Selects ruler marks instead of full-screen graph divider lines.

Tokens: Adds a token to waveform plots. The token also appears on the corresponding axis scale.

X-axis grids: Draws horizontal grids or ruler marks.

Y-axis grids: Draws vertical grids or ruler marks.

Minor grids: Adds minor graph grids between major axis grid markings.

Scope

This option allows you to examine the simulation curves in more detail. After you run the simulation, select the *Scope* option from the AC menu, or press F8. The resulting scope display, for the DIFFAMP example, is shown in Figure 56.

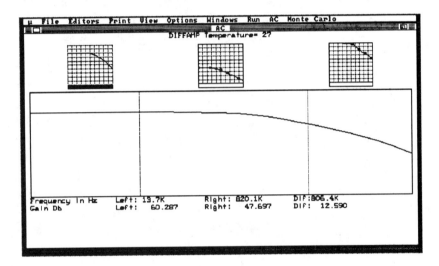

Figure 56

The top portion of the screen shows a plot icon of each of the three curves. The bottom portion shows a blowup of one or more of these waveforms. Initially, the first waveform is displayed, and the lower section of that icon is highlighted. You can select or add other waveforms by clicking on the lower section of the appropriate icon. Alternatively, if you wish to remove a particular curve from the bottom display, you click on the appropriate (highlighted) bar at the bottom of the corresponding icon.

The lower plot has two vertical cursors. These can be used as markers to read values from the curve. In Figure 56, we have set the left cursor (the dotted vertical line) at the flat portion of the amplitude waveform, and the right cursor is at a point 12.59 dB down from this. The value of these cursors is that the X and Y numerical values are printed below the graph. For example, the flat portion of the gain was measured at 13.7 kHz and a value of 60.287 dB. The point on the rolloff occurs at a frequency of 820.1 kHz and a value of 47.697 dB. The printout also presents the difference between these two values. The cursor positions are set by clicking the left mouse button at the desired location of the left cursor and the right mouse button at the right cursor.

KEYBOARD You can use the < and > keys to move the selected cursor left and right. Use the Tab key to toggle between the left and right cursor.

You can also zoom in on portions of the curve. You define a box on the icon by positioning the mouse arrow at the upper left of the desired box and clicking the left button. Then move to the lower right, and press the right button. Alternatively, you can use the + and – keys to resize the box and the Shift+cursor keys to relocate the box. The defined box fills the lower screen. Try this to see how it works.

The portion of the curve you wish to view may be outside the portion you have chosen to plot. You can use the cursor keys to scroll the plot.

Stepping

Component parameters may be stepped from one value to another, producing multiple runs with multiple curves or plots. You activate this feature by selecting *Stepping* from the analysis menu and then entering instructions in the dialog box, which is shown in Figure 57.

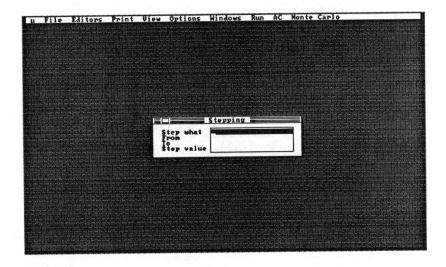

Figure 57

Step What

This is where you specify which parameter you wish to change. For elements, you would enter the name of the element. For example, if you named one or more resistors R1, you would enter **R1**. For devices, you enter the device name, a period, and the parameter you wish to vary. For example, if you used a 2N2222A transistor and wish to vary forward beta, you would type **2N2222A.BF**.

Limits

From specifies the starting value of the parameter. *To* specifies the ending value of the parameter. *Step value* specifies the amount to step the parameter by. As an example, suppose we use our DIFFAMP example circuit and vary transistor Q1 beta from 50 to 250 in steps of 50. That is, we require five runs with betas of 50, 100, 150, 200, and 250. We would select the *AC* option and then select the *Stepping* menu entry. We type in these values:

Step what	Q1.BF
From	50
To	250
Step value	50

Then activate the run either by pulling down the AC menu and selecting *Run* or by pressing F2. The resulting simulation is shown in Figure 58. We have used the Option window to eliminate the phase and group delay plots (recall that these are optional plots) so as not to clutter the figure.

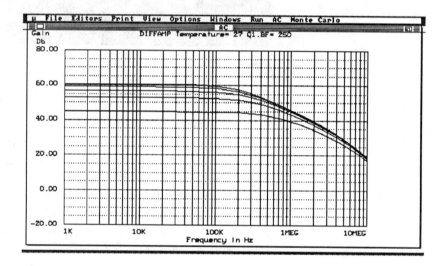

Figure 58

Saving and Printing the Simulation Results

In the section "Run Options," we discussed the three run options. One of these, *Save*, saves the entire simulation for later retrieval. If you exercise this option, and then run the simulation at a later time, you can vary the analysis limits. However, since the program will not rerun the actual simulation, the new analysis limits must be a subset of the original limits. For example, if the Fmax for the original simulation was 1 MHz, subsequent recalls should use a maximum frequency equal to or less than this value.

When you exercise the *Save* option, all node voltages are saved at every frequency point, so this requires a lot of disk space. If you run out of space while saving, an error message will warn you and terminate the run. The portion that was saved is still available for later review.

When you retrieve a saved simulation, you can plot any variable, even if it is one that was not plotted during the initial run. The *Save* option is also available for stepping options, but of course, the memory requirements are even greater.

You must save the circuit from the File menu after the run if you wish to recall the analysis in a later session and have the same analysis limits and options you used during the save run.

Printing the analysis output is simple using the Print pulldown menu. We discussed this menu in the section "Print Menu" in Tutorial 1. Your options are either to print the front window (which would print just the graphics image) or to print the entire screen (which includes everything you see, including the menu bar at the top of the screen). You can also print either of these in a smaller scale using Alt+F3 (for front window) or Alt+F2 (for entire screen).

If your printer does strange things, you might not have set the program for the proper printer. Refer to the section "Options Menu" in Tutorial 1 for the *Graphic setup* option.

Examples

1. Find the low-frequency cutoff for the amplifier shown in Figure 59, where the β of the transistor is given by 200.

   Solution: We input the circuit in the usual manner. Since the beta of the transistor is given as 200, we can either scan the library for an entry with that value or modify the model for any entry in the library. We have chosen the Q0 transistor and set its forward beta to 200.

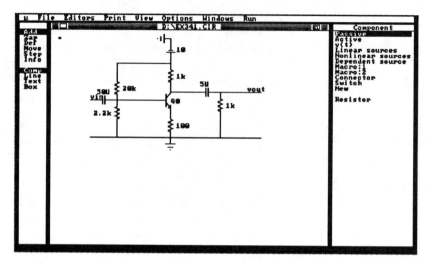

Figure 59

We run the AC simulation to obtain the result shown in Figure 60. We have used the Option menu to specify that only the amplitude appears in the plot. Make sure to identify the input and output node numbers properly. We use the *Scope* option to find the value of gain in the passband and the point at which the gain drops by 6 dB. We find a passband gain of 12.181 dB. The gain drops to 6.181 dB at about 10 Hz.

2. A CD amplifier is shown in Figure 61. Determine the low-frequency cutoff of this amplifier.

   Solution: We input the circuit with a standard NJFET, the 2N4091. We then run the simulation, adjusting the vertical and horizontal axis limits to include the passband and the rolloff. The result is shown in Figure 62. Using the *Scope* option in the AC menu, we find that the gain in the passband is –4.2 dB, and the gain reduces by 6 dB to –10.2 at a frequency of 1900 Hz.

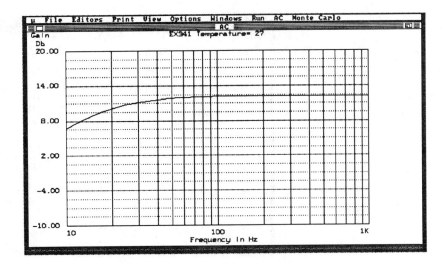

Figure 60

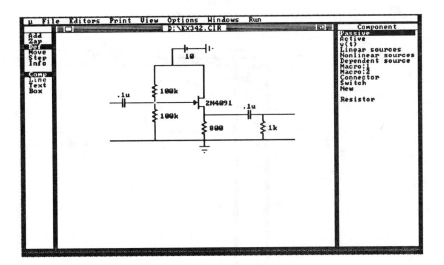

Figure 61

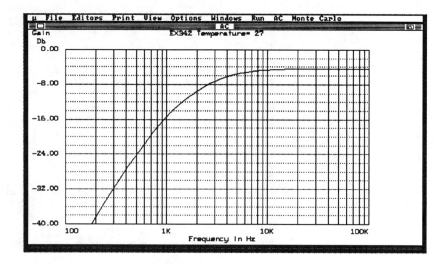

Figure 62

3. Describe the frequency response of a system with transfer function,

$$H(s) = \frac{3s^2 + 60}{s^2 + s + 20}$$

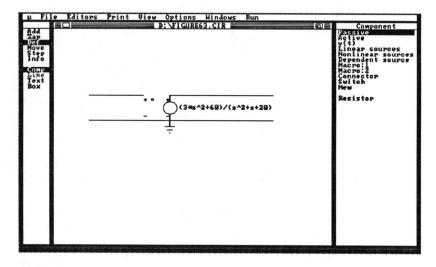

Figure 63

Solution: We use the Laplace source (see the section "Linear Sources") and the circuit shown in Figure 63.

We obtain the response curve of Figure 64.

We have chosen a linear horizontal axis to emphasize that the response is approximately symmetrical around a frequency of 0.71 Hz. This is a notch filter. The result agrees closely with theory since the numerator goes to zero at a frequency of $\sqrt{20}$ radians/sec, which is approximately 0.71 Hz.

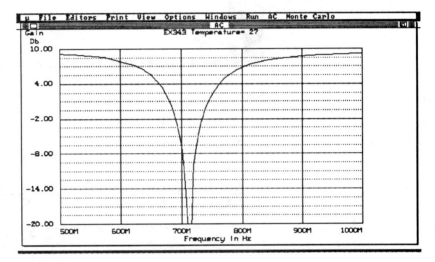

Figure 64

4. You are designing a first-order active filter with a DC gain of 10 and a corner frequency of 1 kHz. The circuit is shown in Figure 65, where you must choose the value of C. Find the value of C using theory. Then use the simulation for that value, stepping on either side of it. Refine the selection with smaller steps until you find the value of C that achieves the specified corner frequency.

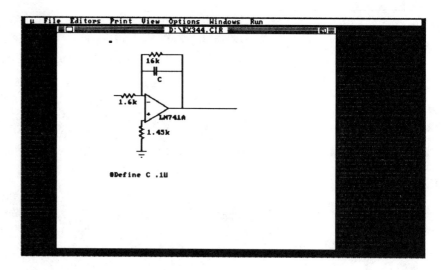

Figure 65

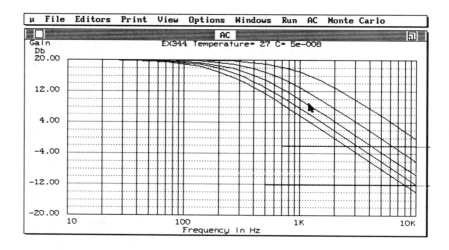

Figure 66

Solution: The theory indicates an approximate value of C to be 0.015 µF. Running the simulation, we step from 0.01 to 0.05 µF to obtain the result of Figure 66. Then we

narrow the range to step from 0.01 to 0.02 µF as shown in
Figure 67. The approximate result is a value of C of 0.016
µF.

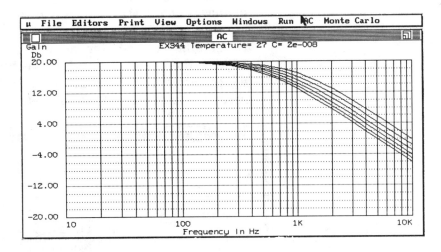

Figure 67

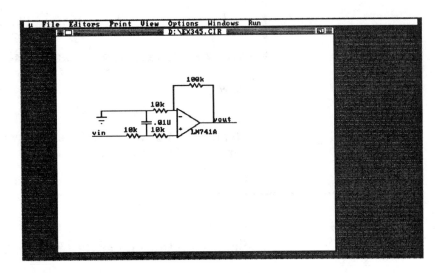

Figure 68

5. What is the DC gain and the corner frequency of the circuit shown in Figure 68?

   Solution: We input the circuit as shown. The result is shown in Figure 69. The DC gain is 11, and the corner frequency is at 2.7 kHz. This is a lowpass filter. You must exercise caution in finding the gain in the passband. When the simulation is run for frequencies between 100 Hz and 10 kHz, the gain value at the lowest frequency is 20.8 dB, which corresponds to a gain of 10.23. However, if we rerun the simulation for a frequency range from 0 Hz to 1

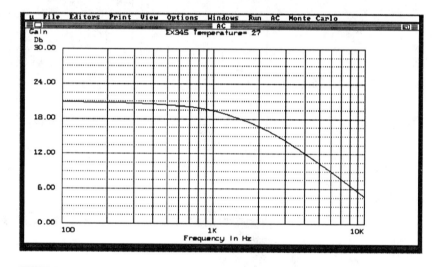

Figure 69

Hz, the gain at the lowest frequency is 20.826 dB, which corresponds to a gain of 11.

Problems

1. Simulate a third-order Butterworth bandpass filter with

$$H(s) = \frac{s^3}{s^6 + 2s^5 + 5s^4 + 5s^3 + 5s^2 + 2s + 1}$$

Plot the amplitude characteristic of this filter. What is the center frequency? Does this agree with the theory?

2. Plot the Nyquist diagram for the feedback system shown in Figure 70. The gain function of the amplifier is given by

$$A_d(s) = (19.2 \times 10^3)\left(\frac{s/0.1}{1+s/0.1}\right)\left(\frac{s/0.5}{1+s/0.5}\right)\left(\frac{s/5000}{1+s/5000}\right)$$

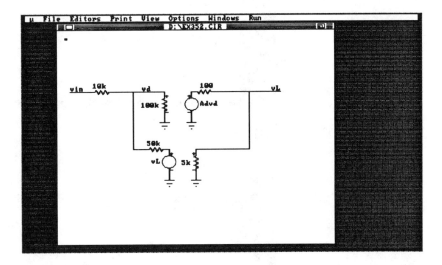

Figure 70

3. An amplifier is shown in Figure 71. Determine the size of the two capacitors, C1 and C2, that will yield a low-frequency cutoff of 20 Hz.

4. Find the high-frequency cutoff for the 2N3904 transistor. Assume that the CE amplifier of Figure 72 is used with the emitter resistor bypassed.

5. A JFET amplifier is designed for a voltage gain of –10 and an input resistance of 50 kΩ. The circuit is shown in Figure 73. Select the capacitor values for a lower frequency cutoff of 20 Hz.

6. Examine the stability of a system with the following open-loop transfer function:

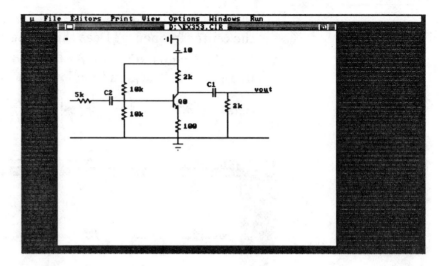

Figure 71

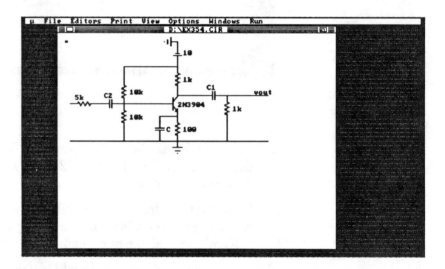

Figure 72

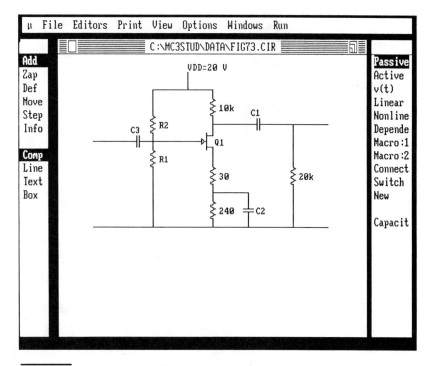

Figure 73

$$G(s)\,H(s) = \frac{A}{s(s+1)\,(0.5s+1)}$$

Do this by plotting the amplitude and phase and by finding the value of A that causes the gain to reach unity when the phase is −180°.

7. An RC high-pass filter is shown in Figure 74. Design this filter for a high-frequency gain of 10 and a corner frequency of 1 kHz.

8. An active filter has a transfer function given by

$$H(s) = \frac{3s - 10}{s + 100}$$

What kind of filter is this?

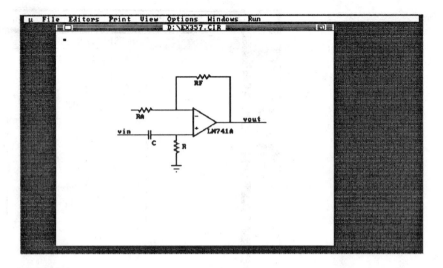

Figure 74

9. Plot a Nyquist diagram for the circuit of Figure 75. To view the critical parts of the diagram, you will want to set the ranges of the real and imaginary parts to several microvolts.

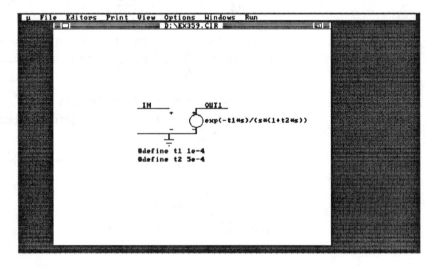

Figure 75

4

DC Analysis

Introduction

DC analysis plots input/output characteristics in the DC condition. The input can be a voltage appearing on a user-specified node (relative to ground) or a differential voltage between two nodes. Alternatively, the input can be a current source. The program evaluates the DC output, either voltage (for a node relative to ground or differential between two nodes), or the current flowing through a resistor specified by the two resistor nodes. The system replaces all inductors with 0.001 Ω resistors and all capacitors with resistors of value 1/GMIN. Recall that GMIN is the minimum conductance value specified in the Global settings menu. It then applies a stepped DC source (voltage or current) to the input and calculates the resulting DC output (voltage or current).

You begin by drawing a circuit, either by entering each component as described in Tutorial 1 or by retrieving a network from the data file. We have been using the differential amplifier circuit for our examples. However, the DC response of this circuit does not provide a great deal of intuitive reinforcement of concepts. Once we finish our example, you can run the differential amplifier DC analysis and see if you can explain the results.

We use the DC analysis simulation to plot transistor operating curves. We select the 2N2222A npn transistor for this example. We have stored a simple testing circuit on your disk under the name CURVES. Recall this file from memory by using the File pull-down menu, selecting *Load Circuit*, and then selecting CURVES from the list of circuits on the disk. The screen should look like Figure 76.

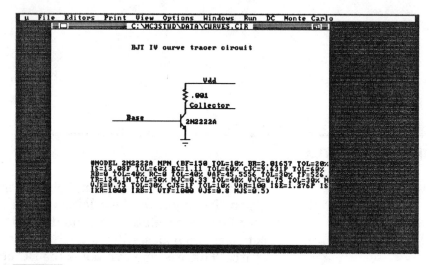

Figure 76

We have labeled the base and collector of the transistor so that we can plot voltages or currents at these points. We have provided a 0.001 Ω resistor in the collector loop to monitor current. We would like to develop the standard operating curves for this transistor, plotting collector current versus collector voltage (the example actually plots Vdd, but since the voltage across the small resistor is negligible, this is essentially collector voltage). We wish to develop the standard parametric curves, where the base current is varied. We choose to develop ten operating curves for base currents between 0 ma and 5 ma in steps of 0.5 ma.

Before running this circuit, you should change the transistor model from the Ebers-Moll to the Gummel-Poon. Do this using the main Options menu and selecting the *Models* portion of this menu. Once you finish this tutorial, you should return the models to the Ebers-Moll (since your simulations will run faster). You might wish to rerun the example using the Ebers-Moll and then try to explain the change in the results based on the differences in the model (the collector currents increase considerably). As we have mentioned, the Gummel-Poon model is more accurate, but the Ebers-Moll is usually sufficient and, because of its simpler configuration, runs faster.

Once you have the network on the screen, you initiate the DC analysis by pulling down the Run menu and selecting *DC analysis*.

KEYBOARD   Type **R** for Run followed by **3** (the number for DC).

Running the DC analysis produces the DC analysis limit window. We discuss this in the next section.

To prepare you for the remainder of this tutorial, we now present a table of contents.

Analysis Limits

After activating the DC analysis, you are presented with a window showing eight limits used in the analysis. When you store a circuit, the limits are stored with it. If you create a new circuit, default limits are used. The limits stored in the CURVES circuit are shown in Figure 77.

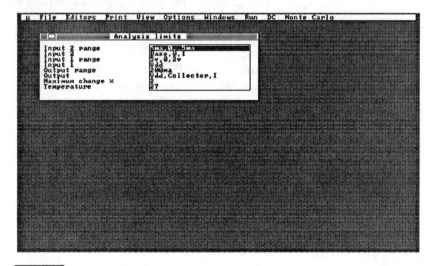

Figure 77

We take the time now to describe each of these analysis limits. However, if you are impatient to see an analysis run, simply pull down the DC menu, and select *Run* (or press F2).

Input 2 range: The program has the capability of stepping two independent inputs. The format is

Final[,Initial[,Step]]

The analysis starts at Initial and ends at Final. The system calculates N solutions spaced equally over the range, Final–Initial with spacing of Step. If you omit Step, it defaults to 0.1*(Final–Initial). If you omit Initial, it defaults to 0.0.

Input 2: This determines where this input will be placed. The format is

Plus[,Minus][,I or V]

Plus is the name of the node where the positive lead of the input source is to be connected. For current sources, it can be thought of as the node to which the directional arrow is pointing. Minus is the name of the node where the negative lead of the input source is to be connected. The third field designates whether the input is a current source or a voltage source. The default for the third field is V (voltage source), and the default for Minus is ground.

If this second independent source is not used, enter **NONE**.

Input 1 range: This is the range of the main input source. The format is the same as that for Input 2 range.

Input 1: This determines where this main input will be placed. The format is the same as that for Input 2.

Output range: This is the range for plotting the output. The format is the same as that for the two input ranges given above.

Output: This field determines where the output will be measured. The format is

Plus[,Minus][,I or V]

Plus is the node name of the positive output lead, and Minus is the node name of the negative output lead. The third field designates whether the output is a current or a voltage. The default for the third field is V, and the default for Minus is ground.

Maximum change %: As the simulation proceeds, the program automatically adjusts the Input 1 step up or down. If the plot changes by more than this percent of full scale, the system reduces the step size. Typical values needed to produce smooth curves range from 1% to 5%.

Temperature: One or more temperature values can be specified for the analysis. The format is

Final[,Initial[,Step]]

The simulation will be performed at temperatures between the Initial and Final, spaced by Step. Thus, for example, an input of 35,20,5 would produce separate runs at 20, 25, 30, and 35 degrees centigrade. If Step is omitted, two runs are performed, one at the Initial and one at the Final. Thus the default value of Step is Final–Initial. If both the Step and Initial are omitted, one run is performed at the Final temperature. Thus, the default value of Initial is Final.

If we examine Figure 77, we see that the selected limits for this example are as follows:

The second input is the current between the base and ground, and we wish to step this from 0 ma to 5 ma, in increments of 0.5 ma.

The main input is the voltage between Vdd and ground, and we wish to step this between 0 volt and 5 volts, in increments of 2 volts. The resulting curves would be essentially the same if we had simply typed **5** for the range. This would range Vdd from 0 to 5 at the default step of 0.5 volt. However, by setting the step at 2 v, the simulation runs faster during those parts of the curve that are relatively flat. Since we have set the maximum percentage change at 2%, the step size decreases considerably below 2 v when the curve slope increases. You should experiment with various values of step value. The optimum choice depends on your choice of maximum percent change.

The output is the current between Vdd and the collector, and this ranges between 0 ma and 200 ma.

The maximum percent change is 2%, and the temperature is room temperature (27 degrees).

If you wish to change any of the analysis limits, click the mouse on that limit, and then type in the new limit.

KEYBOARD To change an analysis limit using the keyboard, position the cursor using the cursor arrow keys, and then edit the entry.

Once you are satisfied with the analysis limits, you are ready to run the simulation. Let us first run the simulation using the variables already selected for plotting. Run the analysis by pulling down the DC menu and clicking *Run*. You can also press F2. The DC analysis is then produced, as shown in Figure 78.

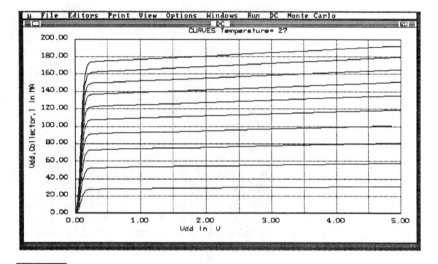

Figure 78

We now analyze the plotting parameters in more detail.

DC Analysis Menu Selections

Once you select *DC* analysis from the Run menu, in addition to presenting the analysis limits, two pull-down menus are added to the top of the screen. These are the DC and the Monte Carlo analysis. We describe the entries in the DC menu in this section. Monte Carlo analysis is not supported in the Student Edition.

Run   Selecting this menu entry runs the simulation.

Limits   This menu entry is checked. It produces the analysis limits if you wish to change them later.

Options   Clicking this menu entry presents a window, as shown in Figure 79.

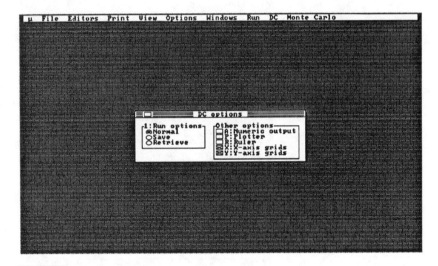

Figure 79

The window is divided into two areas: *Run options* and *Other options*. Options may be selected by clicking the mouse. You can leave the DC options window either by initiating the run (pull down the DC menu and select *Run*) or by erasing the window by clicking on the box to the left of the title. You can also initiate the run by pressing F2.

KEYBOARD   Select options using either the Tab key or the number or letter adjacent to the desired item or group. To select an unnumbered or unlettered option in a group, use the up and down arrow keys to select the option and the space bar to activate the selection. Use the Esc key to exit the DC options window, or press F2 to start a simulation run.

Run Options

Normal produces a normal run. *Save* saves the entire simulation for later retrieval. *Retrieve* retrieves a saved simulation for review.

Other Options

Numeric output: Causes selected waveforms to be sent to a disk file, the screen, or the printer, according to the selections under Text setup. If you have selected a screen output, following display of the curves, you are presented with a table of the type shown in Figure 80.

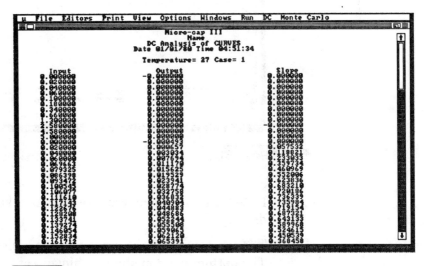

Figure 80

The destination file is named CIRCUITNAME.DNO. (DNO stands for DC Numeric Output.)

Plotter: Not available in the Student Edition.

Ruler: Selects ruler marks instead of full-screen graph divider lines.

X-axis grids: Draws horizontal grids or ruler marks.

Y-axis grids: Draws vertical grids or ruler marks.

Scope

This option allows you to examine the simulation curves in more detail. After you run the simulation, select the *Scope* option from the DC menu, or press F8. The resulting scope display, for the CURVES example, is shown in Figure 81.

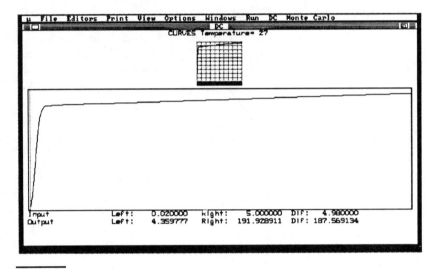

Figure 81

Note that only the last curve (the final step of the Input) is displayed.

The top portion of the screen shows a plot icon of the curve. The bottom portion shows a blowup of the curve. The various control functions are identical to those of the *Scope* option in AC and transient analysis. See the section "Scope" in Tutorials 2 and 3 for details. The two cursors are available for marking points on the curve.

Stepping

Component parameters may be stepped from one value to another, producing multiple runs. You activate this feature by selecting *Stepping* from the analysis menu and then entering instructions in the dialog box, which is shown in Figure 82.

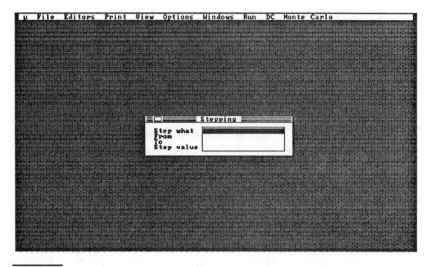

Figure 82

Step What

This is where you specify which parameter you wish to change. For elements, you would enter the name of the element. For example, if you named one or more resistors R1, you would enter **R1**. For devices, you enter the device name, a period, and the parameter you wish to vary. For example, if you used a 2N2222A transistor and wish to vary forward beta, you would type **2N2222A.BF**.

Limits

From specifies the starting value of the parameter. *To* specifies the ending value of the parameter. *Step value* specifies the amount to step the parameter by. As an example, suppose we use our DIFFAMP example circuit and vary transistor Q1 beta from 50 to 250 in steps of 50. That is, we require five runs with betas of 50, 100, 150, 200, and 250. We would select the *Transient* option and then select the *Stepping* menu entry. We type in these values:

Step what	2N2222A.BF
From	50
To	250
Step value	50

Rather than clutter the graph with ten parametric curves for each of five values of beta, let us select the curve for an input current of 3 ma. We therefore change the appropriate limit entry to 3 ma, 3 ma. It is necessary to specify the starting value of 3 ma since a single entry of 3 ma would set the start at a default of 0 and the step at a default of 0.3 ma.

Now activate the run either by pulling down the DC menu and selecting *Run* or by pressing F2. The resulting simulation is shown in Figure 83.

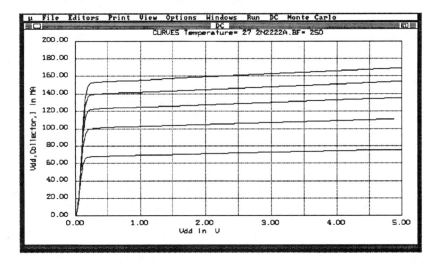

Figure 83

Saving and Printing the Simulation Results

In the section "Run Options," we discussed the three run options. One of these, *Save*, saves the entire simulation for later retrieval. If you exercise this option, and then run the simulation at a later time, you can vary the analysis limits. However, since the program will not rerun the actual simulation, the new analysis limits must be a subset of the original limits. For example, if the Final for the original simulation Input 1 was 10 volts, subsequent recalls should use a maximum voltage equal to or less than this value.

When you exercise the *Save* option, all node voltages are saved at every input value, so this requires a lot of disk space. If you run out of space while saving, an error message will warn you and terminate the run. The portion that was saved is still available for later review.

When you retrieve a saved simulation, you can plot any variable, even if it is one that was not plotted during the initial run. The *Save* option is also available for stepping options, but of course, the memory requirements are even greater.

You must save the circuit from the File menu after the run if you wish to recall the analysis in a later session and have the same analysis limits and options you used during the save run.

Printing the analysis output is simple using the Print pulldown menu. We discussed this menu in the section "Print Menu" in Tutorial 1. Your options are either to print the front window (which would print just the graphics image) or to print the entire screen (which includes everything you see, including the menu bar at the top of the screen). You can also print either of these in a smaller scale using Alt+F3 (for front window) or Alt+F2 (for entire screen).

If your printer does strange things, you might not have set the program for the proper printer. Refer to the section "Options Menu" in Tutorial 1 for the *Graphic setup* option.

Examples

1. You are given the nonlinear circuit shown in Figure 84. Find the input/output characteristic of this circuit.

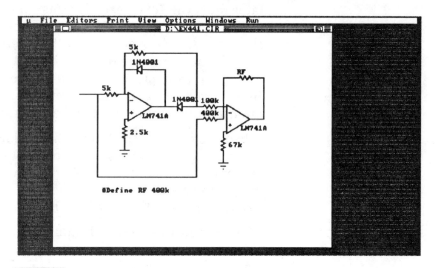

Figure 84

Solution: We run the DC simulation to plot the input/output relationship. After appropriate adjustment of the ranges of the variables, we obtain the graph of Figure 85. This shows that the circuit is a full-wave rectifier, with a gain of approximately 1 for negative inputs and 3 for positive inputs.

2. Find the characteristics of the nonlinear circuit shown in Figure 86.

Solution: We run the DC simulation to plot the input/output relationship. After appropriate adjustment of the ranges of the variables, we obtain the graph of Figure 87. This shows that the circuit is a half-wave rectifier, with output of zero for positive inputs, and −vin for negative inputs.

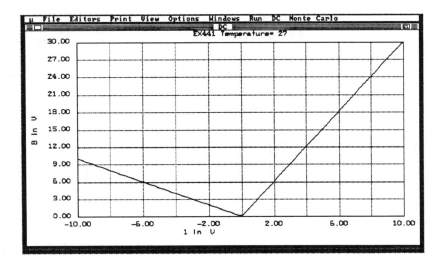

Figure 85

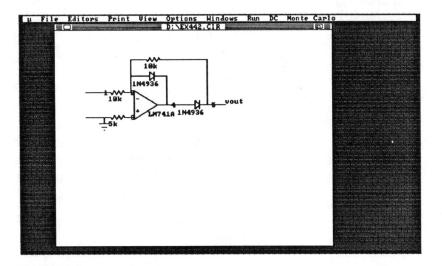

Figure 86

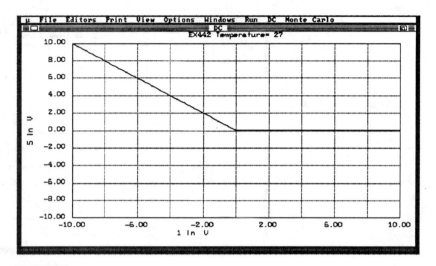

Figure 87

Problems

1. One of the data files supplied with your program is labeled F4. This is a simple circuit, including a vacuum triode. The circuit is configured to plot the relationship between output current and plate voltage while the grid voltage is varied.

   Perform a DC simulation to plot the family of curves. The limits stored on your disk should produce a representative set of curves. Also perform an AC simulation to determine the frequency characteristics of this device.

2. One of the data files supplied with your program is labeled GUMMEL. This simple circuit allows you to examine the nonlinear aspects of the Gummel-Poon transistor model. Access the *Models* option within the Options pull-down menu, and change the transistor model to Gummel-Poon. Then run both a DC and a transient analysis using the limits stored on the disk.

   Examine the family of curves produced by the DC analysis, and try to determine beta from these curves.

Then examine the transient analysis to see the dependency of beta on the collector current. Note that β goes to zero for both high and low collector currents. Also note that we have modified the Transient monitor menu to plot the log of the current variables.

3. What is the function of the circuit shown in Figure 88?

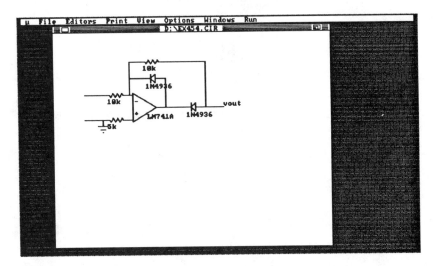

Figure 88

4. What is the function of the circuit shown in Figure 89?

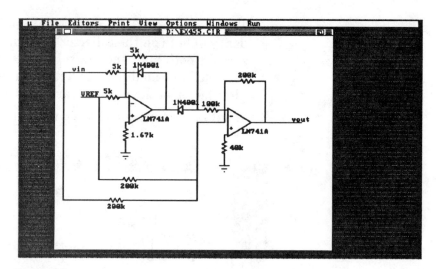

Figure 89

5

Fourier Analysis

Introduction

Fourier analysis consists of evaluating an infinite series of sine and cosine terms that approximates a periodic function, in this case, a waveform or voltage as a function of time.

The Fourier analysis option allows you to perform a Fourier analysis on a waveform that is stored on disk in the form of 256 successive time samples. The program can store up to ten waveforms in a directory, in files named USER.0 to USER.9.

The waveform being analyzed may be generated by a transient analysis or by external means. The waveform is simply a collection of N data points stored in a waveform file. The waveform text file structure is (where we designate x(t) as the waveform):

N

x(0)

x(DT)

x(2DT)

.

.

.

x[(N−1)DT]

N is the number of data points in the file, and DT is the spacing between samples. The program does not need to know this sampling period since the results are normalized in frequency. The program is finding the Fourier series of a time-limited waveform. This can be considered a sampled version of the Fourier transform. The first harmonic is at a frequency that is the reciprocal of the period of the input waveform. The period of this waveform is NDT.

Saving a Waveform

The simplest way of creating a waveform text file is from the transient analysis program. You can create any waveform you wish by designing a simple circuit containing a source that generates the desired waveform (nonlinear or linear). One of the menu items in the Transient analysis pull-down menu is *Monitor*, and if you select that entry, you are presented with a monitor table that looks like Figure 90, which resembles Figure 34.

We have made one important change, however. We have checked the *Ur* column for the V(x,y) voltage output. This causes that output to be saved in a waveform user file under the name DIFFAMP.USR. (You can check only one waveform for storage in the user file. Thus there is no ambiguity in the name of the file.) The waveform is always stored in a file

Figure 90

with the circuit name followed by a USR extension. The number of data points, N, stored in this file depends on the transient relative error, which is set in the Transient limits window (see the section "Analysis Limits" in Tutorial 2). The number of points is a power of 2 and varies according to the following table. N = 128 for a relative error of 1%, and N doubles for every decrease in error by a factor of 10.

Relative error	N
1E-2	128
1E-3	256
1E-4	512
1E-5	1024
1E-6	2048
1E-7	4096

The file is in the form specified in the section "Introduction," so it is ready for Fourier analysis.

Fourier Analysis

The Fourier analysis is initiated from the Run pull-down menu. Simply click on this entry, or type **R** (for Run) followed by **4**.

You are presented with a window showing the .USR files that have been stored in the data subdirectory. The first several entries allow you to log on to a different directory.

If you have not stored a waveform in a user file as described in the previous section, the only file you will see is one called SAMPLE. This was shipped with the program. Select this file either by clicking on it or by highlighting it and pressing ↵.

The system then requests the number of harmonics you wish to calculate. This is related to resolution, and you would usually choose the highest number (128) unless your waveform is periodic. Select the number of points by highlighting it and then pressing ↵.

The system then reads the waveform file, performs a fast Fourier transform, and plots the waveform. The result is shown in Figure 91.

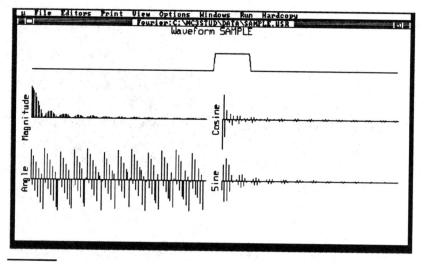

Figure 91

The waveform is plotted on the top. This is followed by four waveforms (actually line drawings with a line for each of the 128 harmonics). Since the Fourier series consists of sines and cosines, there are two ways to present this. The right side of the graph shows the amplitudes of the cosine and sine terms at each harmonic frequency.

The sines and cosines can be combined using trigonometric identities, as follows:

$$A\cos(\omega t) + B\sin(\omega t) = C\cos(\omega + \theta)$$

where

$$C = \sqrt{A^2 + B^2}$$

$$\theta = -\tan^{-1}\left(\frac{B}{A}\right)$$

The magnitude, C, and the phase angle, θ, are plotted on the left side of the screen.

Hardcopy

When you run a Fourier analysis, the Hardcopy pull-down menu appears at the top of your screen. When you pull down this menu, you note that it has three entries:

1:Text Output

2:Plotter

3:Palette

Plotter and Palette are not supported in the Student Edition of MICRO-CAP III.

If you select *Text output*, the output waveforms, in tabular form, are sent to the screen, a disk file, or the printer depending on how you set the global *Text setup* option (see the section "Options Menu" in Tutorial 1). The first page of the printout for this example is shown in Figure 92.

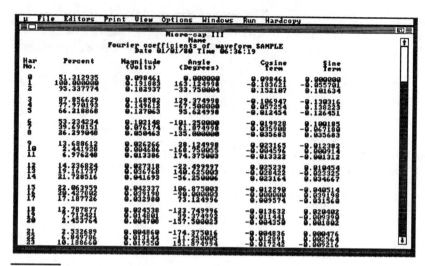

Figure 92

For each harmonic, an entry shows the height of the graph.
There is an additional column labeled *Percent*. This is the
ratio of the magnitude to the maximum magnitude, in percent. Also note that the 0-harmonic magnitude cosine term is
the average value of the waveform.

Problems

1. Devise an electronic circuit to produce an approximation
   to an impulse at the output. (*Hint*: Consider an operational amplifier configured to differentiate the pulse input.)
   Then find the Fourier transform of the impulse.

2. Use the approximate impulse of Problem 1 as the input to
   an RC circuit. Plot the output, and find its Fourier transform. Compare this with the amplitude and phase plot of
   the RC circuit as obtained using the AC analysis simulation. Explain any discrepancies.

III

REFERENCE

1

Glossary

The following is an alphabetical listing of key functions and terms used in the Student Edition of MICRO-CAP III. Following each entry in brackets is the page number where you will find additional information.

AC analysis [101] One of the types of simulation performed by MICRO-CAP, this is used to plot the small-signal transfer characteristics of a circuit.

AC menu [107] Once you select *AC* from the Run menu, the AC menu is added to the pull-down menus at the top of the screen.

Active [37] Clicking this entry on the Components menu allows you to select active components, which includes op-amps, bipolar transistors, MOS transistors, and FET transistors.

Add [33] The mode (from the window on the left side) that permits you to add an element to the existing circuit.

Analysis limits [71, 104, 130] A window presented after selecting a type of simulation from the Run menu. It is used to set limits on the simulation.

Analysis options [75] A selection in the Options window for transient analysis, this window allows you to select one of four modes of analysis that affect the simulation and convergence.

Battery [41]   Part of the v(t) components menu selection, this is a DC voltage source.

Box [53]   Part of the mode window at the left of the screen. Allows you to define a box for repetition (step) or deletion (zap). With the mode selections of *Def/Box*, you use the left button to mark the upper-left corner and right button to mark the lower-right corner of the box.

Calculator [66]   One of the entries in the μ pull-down menu, this puts the program into an expression-type calculator.

Change data path [62]   An entry in the File pull-down menu that allows you to alter the data path for library, circuit, and output files.

Click [6]   The process of pressing and releasing a button on the mouse.

Click and drag [7]   The process of pressing a button on the mouse, then moving the mouse while holding the button down. This is useful in positioning components or in pulling down menus from the top of the screen.

Comp [33]   The selection (from the window on the left side) that tells the program you are working with components. Depending on your selection from the Mode window, you may be adding, deleting, moving, or modifying the value of a component.

Component [56]   One of the libraries accessed from the Editor pull-down menu. This editor controls the name, shape, parameter text location, electrical definition, and pin assignments for each component in the library.

Cross [39]   Part of the Connections menu, this is used when two wires cross where you do not wish the two wires connected to each other. The cross connects the two line segments on either side of the second line (at right angles).

Cursor [7]   The flashing object on the display.

Connections [39]   An entry in the Components menu that allows you to select grounds, shorts, jumpers, ties, and crosses for addition to your circuit.

Create new circuit [61]   An entry in the File pull-down menu that clears the screen so that you can start a new circuit.

Data path [13]

The place the program looks for data and writes circuits and results to. Normally, C:\MC3S\DATA for hard disks and B:\DATA for floppy disks.

DC analysis [127]

One of the types of simulation performed by MICRO-CAP, this is used to plot input/output characteristics in the DC condition.

DC menu [133]

Once you select *DC* from the Run menu, the DC menu is added to the pull down menus at the top of the screen.

DC relative error [64]

One of the entries on the *Global settings* selection of the Options pull-down menu. This is used in solving nonlinear DC equations to determine when the iterative process should stop. Typical values are between 1E-3 and 1E-6.

Def [54]

Selected from the Mode window on the left of the screen. In the *Box* mode, this is used to define the box. In other modes, it is used to change a component parameter or text. You click on the item to be changed and are then prompted to type the new material.

#Define statement [36]

Text material added to the diagram to define the value of a passive component. May contain mathematical operations.

Delay range [106]

One of the entries in the Analysis limits window for AC analysis selected from the AC analysis pull-down menu. This is used to set the range for the scale of delays.

Dependent source [48]

One of the selections in the Components menu, this allows you to add a source where the source parameters are proportional to the controlling parameters. You are prompted to enter the proportionality factor.

Devices [56]

One of the libraries accessed from the Editor pull-down menu. This contains the models of op-amps; diodes; bipolar transistors; MOSFETs; JFETs; and the pulse, sinusoidal, and polynomial sources.

Drawing elements [34]

Once the *Add* mode has been selected and you have chosen the component from the components menu, you draw elements by positioning the mouse arrow at the desired location and clicking the left mouse button. You can click and drag to position the element, and you can click the right button (while holding the left) to rotate the element.

Ebers-Moll model [64]	One of two possible models for bipolar transistors that you select from the *Models* selection in the Options pull-down menu. This model is simpler than the Gummel-Poon model since it ignores certain second-order (nonlinear) effects. The simulation runs faster with this model.
Editors menu [56]	One of the pull-down menus at the top of the screen. This accesses the library editors, which contain models of devices and components. By selecting one of the libraries and then selecting the device or component, you can change the entries to customize the elements for your circuit.

Engineering notation [35]   MICRO-CAP III recognizes the following abbreviations:

F	Femto	1E-15
P	Pico	1E-12
N	Nano	1E-9
U	Micro	1E-6
M	Milli	1E-3
K	Kilo	1E3
MEG	Mega	1E6
G	Giga	1E9
T	Tera	1E12

FET transistors [39]	These are included in the *Active component* menu selection. MICRO-CAP III has a library of 49 standard transistors and 2 additional user-defined transistors.
File menu [61]	One of the pull-down menus at the top of the screen that allows you to load, unload, or save circuits or libraries; change data paths; or clear the screen.
Fourier analysis [145]	One of the types of simulation performed by MICRO-CAP, this is used to evaluate the discrete Fourier transform of a waveform stored on the disk.
Frequency range [104]	One of the entries in the Analysis limits window for AC analysis selected from the AC pull-down menu. This specifies the range of frequencies over which the transfer function is evaluated.
Frequency step [109]	Part of the Options window for AC analysis, this allows selection of automatic frequency stepping or fixed step sizes.

Gain range [105]   One of the entries in the Analysis limits window for AC analysis selected from the AC pull-down menu. This sets the limits of the gain axis.

Global settings [64]   Part of the Options pull-down menu. Allows you to set a number of values used in the simulations, including number of iterations, factors relating to iterations and convergence, and switch resistances.

Ground [39]   Part of the Connections menu. Every circuit must have a ground for MICRO-CAP III to perform a simulation.

Gummel-Poon model [64]   One of two possible models for bipolar transistors that you select from the *Models* selection in the Options pull-down menu. This model is more complex than the Ebers-Moll model since it includes certain second-order (nonlinear) effects. The simulation runs slower with this model.

Hardcopy [149]   When you run a Fourier analysis, an additional pull-down menu appears at the top of the screen. Choosing *Text output* from this menu sends the output waveform to the screen, a disk file, or the printer depending on the global text setup.

Help [66]   Accessed either from the μ pull-down menu or by pressing F1. The menu yields a general help window, whereas F1 gives specific help relative to the current configuration.

Horizontal scale [109]   Part of the Options window for AC analysis, this selects whether the horizontal scale should be linear or logarithmic (dB).

Info [55]   Part of the Mode window, this mode provides information on any clicked component.

Initial values [76]   One of the selections in the Options window for transient analysis, this allows you to choose several forms of initial capacitor voltages and inductor currents.

Input [106]   One of the entries in the Analysis limits window for AC analysis selected from the AC pull-down menu. This is used to specify the location of the input source.

Input 1 [131]   One of the selections in the Analysis limits window for DC analysis. This is the first of two inputs that can be stepped.

Input 2 [131]   One of the selections in the Analysis limits window for DC analysis. This is the second of two inputs that can be stepped.

Input 1 range [131] One of the selections in the Analysis limits window for DC analysis. This allows you to specify the stepped values of the first of two inputs.

Input 2 range [130] One of the selections in the Analysis limits window for DC analysis. This allows you to specify the stepped values of the second of two inputs.

Installation on floppy disk [11] Simply copy the two disks to create working files. Then start the program by typing **MC3S**.

Installation on hard disk [12] With original program disk in drive A, type **INSTALL C:\MC3S**.

Isource [41] Part of the v(t) components menu selection. A DC current source. After drawing the source, you are prompted to insert the current value.

Jumper [33] (See **Cross**.)

Limits [75, 108, 134] A selection on the Transient AC and DC pull-down menus, this activates the Limits window.

Line [53] Part of the Mode window. When in this mode, you are adding (or deleting) line segments. In adding, you click the left button on one end of the line and the right button on the other end. Lines must be either horizontal or vertical.

Linear source [44] This is selected from the Components menu and is specified by a Laplace transfer function (in formula or tabular form). You must select one of eight possible controlled source configurations.

Load circuit [61] An entry in the File pull-down menu that allows you to load a circuit file from disk. You are presented with a file dialog box listing all circuits saved in that file. It also allows you to change to a different drive.

Load library [62] An entry in the File pull-down menu that allows you to load a new device library from disk.

Macro [49] Part of the Components menu. A circuit with grid text labels that defines the connections with the pins of the shape in which it is contained. You can define a macro for any small circuit configuration that you use repeatedly.

Main variable [109] Part of the Options window for AC analysis, this allows selection of the variable to be plotted on the main plot.

Maximum change % [73, 106, 131]   One of the entries in the Analysis limits window for transient, AC, and DC analysis. If the calculated value represents a change by more than this percentage, the system reduces the size of the time step.

Maximum current error [74]   One of the entries in the Analysis limits window for transient analysis. This is the absolute error tolerance for current.

Maximum voltage [64]   An entry on the *Global settings* selection of the Options pull-down menu. It affects DC convergence and should be set between 1E3 and 1E6.

Maximum voltage error [74]   One of the entries in the Analysis limits window for transient analysis. This is the absolute error tolerance for voltage.

Merge circuit [62]   An entry in the File pull-down menu that allows you to merge a circuit from disk with that currently in the front window. Merging occurs at the cursor location.

Minimum conductance [64]   An entry on the *Global setting* selection of the Options pull-down menu, this is the minimum conductance of any component. Typical values are from 1E-9 to 1E-15.

Minimum timestep [74]   One of the entries in the Analysis limits window for transient analysis, this is the smallest time step allowed during the analysis.

Models [64]   One of the selections in the Options pull-down menu, this permits you to select models for bipolar transistors, MOSFET devices, and op-amps.

#Model statement [37]   Text material on the diagram that allows you to define parameters of a device, as in the device library.

Monitor [82]   A selection in the Transient pull-down menu, this permits you to specify which waveforms are to be plotted.

MOS transistors [38]   These are included in the *Active component* menu selection. MICRO-CAP III has a library of 41 standard transistors and 10 additional user-defined transistors.

Move [54]   Selected from the Mode window on the left of the screen. To move an element, click and drag the element to the desired location. You can rotate the element by clicking the right button while holding down the left button.

μ menu [66]   One of the pull-down menus at the top of the screen. Among five possible selections are *Calculator*, *Help*, and *Quit*.

Netlist [63]

A tabular description of a circuit listing each element and the nodes (by number) to which it is connected.

New [53]

An entry in the Components menu that allows you to define your own component or macro and add this to the library.

Node number [56]

MICRO-CAP III numbers all nodes, starting with 0 for the ground. You need to know the node numbers to specify which output(s) you want plotted. (See **Show node numbers.**)

Node snap [64]

An entry in the Options pull-down menu. When enabled, component locations are adjusted so that pins coincide with the nearest circuit component that is within two grids. Thus, when you position a component in a circuit, you do not need to fine-tune its location so that wires touch.

Nonlinear source [45]

This is selected from the Components menu, and you must select one of eight possible controlled sources. The source is then specified by a formula or a table.

NPN transistors [38]

These are included in the *Active component* menu selection. MICRO-CAP III has a library of 39 standard transistors and ten additional user-defined transistors.

Number of iterations [64]

An entry on the *Global settings* selection of the Options pull-down menu. This tells the system the maximum number of times to iterate a solution before aborting the analysis.

Op-amps [39]

These are included in the *Active component* menu selection. MICRO-CAP III has a library of 43 standard transistors and two additional user-defined op-amps.

Optional plots [109]

Part of the Options window for AC analysis, this selects whether or not you wish to plot phase angles and group delay in addition to amplitude.

Options menu [63]

One of the pull-down menus at the top of the screen. Allows you to set preferences (see **Preferences**) and global settings (see **Global Settings**) and to choose the type of model for the transistors and op-amps.

Options window [75, 108, 134]

One of the entries in the Transient, AC, and DC pull-down menus, this activates the Options window that controls run options, initial values, analysis options, and other options.

Other options [77, 110, 135]

One of the selections in the Options window for transient analysis, AC analysis, and DC analysis, this allows you to control certain aspects of the output plot. You can send the output to a disk file or printer, you can decide how detailed you want the grid lines in the plot, and you can add tokens to distinguish one curve from another.

Output [106, 131]   One of the entries in the Analysis limit window for AC analysis and DC analysis. This is used to specify the location of the output.

Output range [131]   One of the selections in the Analysis limits window for DC analysis, this specifies the range for plotting the output variable.

Passive [34]   Clicking this entry on the Components menu allows you to select passive elements, which include resistors, diodes, transformers, capacitors, inductors, and lines.

Phase range [106]   One of the entries in the Analysis limits window for AC analysis selected from the AC pull-down menu. This is used to set the scale of the phase angle axis.

PNP transistors [38]   These are included in the *Active component* menu selection. MICRO-CAP III has a library of 39 standard transistors and ten additional user-defined transistors.

Polynomial sources [48, 61]   Part of the *Dependent source* selection from the Components menu. Provides for an input/output relationship that follows a polynomial equation.

Powers of 10   (See **Engineering notation**.)

Preferences [63]   One of the entries in the Options pull-down menu. Allows you to select the following: mouse ratio relating mouse movement to arrow movement on the screen; whether or not sound accompanies warnings and error messages; whether or not a warning appears when a circuit has changed; whether or not the file dialog box appears on the screen; whether or not to seek verification before quitting; whether or not the program checks the circuit for floating nodes; whether or not components are adjusted so that pins coincide with nearest circuit component. (See **Node snap.**)

Print circuit [63]   One of the entries in the Print pull-down menu that allows you to print only the circuit on your printer. If you have selected to view node numbers, these will print with boxes around them.

Print entire screen [63]	One of the entries in the Print pull-down menu that transfers a graphic image of the entire monitor display to the printer. This option was used in many of the figures in this text.
Print front window [63]	One of the entries in the Print pull-down menu that allows you to print the graphics image of the front window. If the front window contains the circuit, this will print a different type of image than the *Print circuit* instruction. Node numbers will print as they are displayed on the monitor.
Print interval [73]	One of the entries in the Analysis limits window in transient analysis. It is used if you are printing a table of outputs in numeric form.
Print menu [63]	One of the pull-down menus at the top of the screen that allows you to print the circuit, the front window, the entire screen, or a netlist to your printer.
Print netlist [63]	One of the entries in the Print pull-down menu. This option creates a netlist showing the contents of the current circuit. It sends this netlist to the destination specified in the *Text setup* option that is part of the Options menu.
Pt [83]	One of the entries in the *Monitor window* selected from the Transient pull-down menu. A mark in this column tells the system to plot the specified waveform on the transient graphs.
Pulse source [41]	Part of the v(t) components menu selection. Includes simple pulse sources as defined in the device library. You can modify the library entries from the device library.
Quit [66]	One of the entries in the μ pull-down menu, this is used to exit MICRO-CAP. It cannot be accessed if a simulation is being run. You must first exit the simulation.
Relative error [73]	One of the entries in the Analysis limits window for transient analysis. This is used in nonlinear circuit analysis. It determines when an iteration is considered to have converged. Typical value of 1E-3.
Relaxation factor [64]	An entry on the *Global setting* selection of the Options pull-down menu. It is used in operating point calculations and can be thought of as a kind of "throttle." It is usually between

w for tran-
e simulation is
ime, and max-

ver values should be used if the simulation is
ng with the higher values.

udes 16 pos-
e device
the device

elections in the Transient, AC, and DC pull-down
initiates the simulation.

pull-down menus at the top of the screen, this per-
o select any of the types of simulations (transient,
or Fourier) or to exit a simulation.

he Options window, this gives you a choice of three
ions: a normal run, one that saves the simulation, or
it retrieves a saved simulation.

you to

try in the File pull-down menu that saves the current
t under the current circuit name and disk file. This will
ce the existing stored circuit with the one currently in
ront window.

or DC pull-
rameter
d Step
pped

entry in the File pull-down menu that saves the current
cuit under a name you supply. You are prompted for the
w filename.

ll-down
runs
value.

n entry in the File pull-down menu that allows you to save
he library in memory to disk under a name you specify.

ree
d, or

One of the selections in the View pull-down menu. Allows
you to display the front circuit at normal scale, 1/2 size, or
1/4 size.

e Op-
tch in

A selection in the Transient, AC, and DC pull-down menus,
this allows you to examine the results of a simulation in
more detail and to read out numerical values for samples of
the curves.

Op-
ch in

Part of the Connections menu. You have a choice of seven different lengths of shorts that can be added to the circuit.
These are simply straight pieces of wire.

rmal-
disks.

kt

One of the entries in the View pull-down menu. When
checked, the component parameter text is displayed.

m

kt [56]

One of the entries in the View pull-down menu. When
checked, the text markings on the grids of the plots are
displayed.

rd

One of the entries in the View pull-down menu. When
checked, the node numbers are displayed on the circuit.

56]

Simulation time [72] — One of the entries in the Analysis limits windo sient analysis, this sets the time over which th performed. You give the starting time, ending t imum time increment used in the iteration.

Sine source [41] — Part of the v(t) components menu selection. Incl sible sinusoidal sources, which are defined in th library. You can modify the library entries from library.

Starting the program [12] — After installation, simply type **MC3S** followed by

Step [54] — Selected from the Mode window, this mode allows repeat portions of the circuit.

Step what [81, 113, 137] — When you select *Stepping* from the Transient, AC, down menus, you are first asked to specify what pa should be stepped. This is followed by *From, To,* an *value*, where you specify the actual values of the ste parameter.

Stepping [80, 112, 136] — One of the selections in the Transient, AC, or DC pu menus, this permits you to make multiple simulatio with varying values of any parameter or component

Switch [51] — Part of the Components menu. You can add any of th types of switches: current-controlled, voltage-controll time-controlled.

Switch on-resistance [64] — One of the entries on the *Global setting* selection of th tions pull-down menu. This sets the resistance of a sw the on state.

Switch off-resistance [64] — One of the entries on the *Global setting* selection of the tions pull-down menu. This sets the resistance of a swit the off state.

Sys path [13] — The place the program looks for system instructions. No ly, C:\MC3S\SYS for hard disks and B:\SYS for floppy

System requirements [3] — You need an IBM PC, XT, AT, PS/2 or a compatible syste with MS-DOS or PC-DOS, version 3.0 or higher. At least 640K of RAM is required, and you must have one of the specified graphics adapters. Two 360K disk drives or a ha disk are required. A mouse is optional but very useful.

Temperature
[73, 106, 132]

One of the entries in the Analysis limits window for transient, AC analysis, and DC analysis. You can specify one or more temperatures and can instruct the simulation to iterate through a number of temperatures to produce a family of curves.

Text [53]

Selected from the Mode window on the left of the screen. You can add (or delete) text from anywhere on the diagram. It can be used to give the viewer information, to label nodes, or as part of #Define or #Model statements.

Tie [39]

Part of the Connection menu, this permits connection of nodes that are separated. You need not connect wires between each pair of nodes. The tie is used to mark the nodes with identical labels. They are then electrically connected.

TO [83]

One of the entries in the Monitor window selected from the Transient pull-down menu. A mark in this column tells the system to print the numeric value of the waveform to a screen window, a disk file, or the printer.

Tokens [77]

A selection in the *Other options* selection for the Options window for transient analysis, you can add a token to waveforms on the graph. This helps distinguish among waveforms, particularly when you are dealing with a noncolor copy.

Transient analysis [69]

One of the types of simulation performed by MICRO-CAP, this is used to plot time waveforms at various points in the circuit.

Transient menu
[75]

A pull-down menu that appears at the top of the screen after you select transient analysis from the Run menu.

Unload circuit [62]

An entry in the File pull-down menu that removes the front circuit window from memory. You need to select *Create new circuit* before drawing a new circuit.

Ur [83]

An entry in the Monitor window that is selected from the Transient pull-down menu. A check mark in this column tells the system to save the waveform to a user file source data file.

User function [42]

Part of the v(t) components menu selection, this is a source that you define by an algebraic expression. We suggest that you use the nonlinear source instead since it can do everything that the User function does (and more), and the User function may be phased out in later versions of MICRO-CAP.

User source [42] Part of the v(t) components menu selection. You must have created an ASCII text file with N successive time samples of the source function.

Vertical scale [109] Part of the Options window for AC analysis, this selects whether the vertical scale should be linear or logarithmic (dB).

View menu [56] One of the pull-down menus at the top of the screen. Allows you to scale circuits, select whether or not to display component parameter text or grid text, and whether or not to display node numbers.

v(t) source [40] Part of the Components menu. Includes six types of voltage sources: battery, pulse source, sine source, I source, user source, and user function.

Windows menu [55] One of the pull-down menus at the top of the screen. You can open or close a window by clicking on the entry in the menu. Windows can also be moved or resized.

Zap [54] Selected from the Mode window. This is used for deleting elements, text, or the contents of a box. Simply enter this mode and click on the item to be deleted.

Zin range [105] One of the entries in the Analysis limits window for AC analysis selected from the AC pull-down menu. This is used to scale the input impedance axis.

Zout range [105] One of the entries in the Analysis limits window for AC analysis selected from the AC pull-down menu. This is used to scale the output impedance axis.

2

Models Used by MICRO-CAP III

In this appendix, we discuss the diode, BJT, and op-amp
models, and we give references where you can obtain further
information on the models used in MICRO-CAP III (and in
SPICE). Our purpose is not to discuss the theory of modeling
but to give you the necessary information to modify existing
models or create your own from information normally available in data books.

Diodes

The diode model used in MICRO-CAP III is the standard exponential version with breakdown effects, similar to that found
in most SPICE programs. The equivalent circuit consists of a
diode with a parallel conductance, *GO*, and a series resistance,
RS. The equations and parameters are as follows:

V	the voltage applied across the diode
VT	the thermal voltage (26 mV at room temperature)
BV	breakdown voltage
RL	leakage resistance
ID	current flowing through the diode
IS	saturation current

If $V < -BV$,

$$ID = -IS(e^{-q(BV + VD)/kt} - 1 + q*BV/kT)$$

If $-BV < V < -5*n*Vt$,

$$ID = \frac{V}{RL} - IS$$

If $V \geq -5*n*Vt$,

$$ID = \frac{V}{RL} + IS*e^{V/(n*VT)} - 1$$

The thermal voltage is computed as follows:

$$VT = k*q/T$$

where

k = Boltzmann constant ($1.38*10^{-23}$ J/K)

q = electron charge ($1.602*10^{-19}$ C)

T = nominal temperature ($300°K$)

The junction capacitance of the diode depends on the voltage.
If $V \leq FC*VJ$, then

$$CJC = \frac{CJO}{\left(1 - \frac{V}{VJ}\right)^M}$$

If $V>FC*VJ$, then

$$CJC = CJO*(1-FC)^{-(1-M)}*\left(1-FC*(1+M)+M*\left(\frac{V}{VJ}\right)\right)$$

The temperature dependence of the saturation current is given by the following equation:

$$IS(T) = IS*e^{(T/Tnom-1)*EG/(N*VT)}*\left(\frac{T}{Tnom}\right)^{XTI/N}$$

where

IS	=	saturation current at $Tnom$
XTI	=	saturation exponent for IS
N	=	emission coefficient
t	=	temperature under simulation
$Tnom$	=	nominal temperature

Bipolar Transistors

Two models of bipolar transistors are available, the Gummel-Poon and the Ebers-Moll. The Ebers-Moll model is mathematically the same as the Gummel-Poon model with certain second-order effects ignored.

The Ebers-Moll model provides a good first-order model of the terminal currents and charge storage effects. The Gummel-Poon model adds base width modulation, high level injection, and base widening effects.

The model contains the following parameters. In this list, BE stands for base-emitter, and BC for base-collector. The parameters marked by an asterisk (*) are not used in the Ebers-Moll model. That is, the system sets them either to zero or to infinity to null out their effects.

Name	Parameter	Units
BF	Maximum Forward β	—
BR	Reverse β	—
XTB	Temperature coefficient for β	—
IS	Saturation current	amps
EG	Energy gap	eV
CJC	BC zero-bias depletion capacitance	farads
CJE	BE zero-bias depletion capacitance	farads
RB	Zero-bias base resistance	ohms
RC	Collector resistance	ohms
*VAF**	Forward Early voltage	volts
TF	Forward transit time	seconds
TR	Reverse transit time	seconds
MJC	BC grading coefficient	—
VJC	BC built-in voltage	volts
MJE	BE grading coefficient	—
VJE	BE built-in voltage	volts
CJS	Collector-substrate 0-bias capacitance	farads
*VAR**	Reverse Early voltage	volts
NF	Forward emission coefficient	—
NR	Reverse emission coefficient	—
*ISE**	BE saturation current	amps
*ISC**	BC saturation current	amps
*IKF**	Corner for Forward β high-current roll-off	amps
*IKR**	Corner for Reverse β high-current roll-off	amps
*NE**	BE leakage emission coefficient	—
*NC**	BC leakage emission coefficient	—
RE	Emitter resistance	ohms
*IRB**	Current where base resistance falls by half	amps
*RBM**	Minimum base resistance at high currents	ohms
*VTF**	*VBC* dependence of *TF*	volts
*ITF**	*IC* dependence of *TF*	amps
*XTF**	Coefficient dependence of *TF*	—
*XCJC**	Fraction of BC depletion capacitance to internal base node	—
VJS	Substrate-junction built-in potential	volts
MJS	Substrate-junction grading coefficient	—
XTI	Saturation current temperature exponent	—
KF	Flicker-noise coefficient	—
AF	Flicker-noise exponent	—
FC	Forward-bias depletion coefficient	—

For details regarding the simulation equations, refer to the references at the end of this appendix. For now, we qualitatively discuss some of the operational aspects of the model.

A number of terms in the above list are temperature dependent. These parameters are automatically computed internally by the program.

BF, the forward beta, varies with temperature according to the temperature coefficient, **XTB.** It also varies with collector current in the Gummel-Poon model and is constant in the Ebers-Moll model. The nominal value is obtained directly from data sheets.

BF, the reverse beta, also depends on temperature and current. This parameter is important only if you are operating the transistor in the reverse-active region.

IS, the saturation current, follows the same relationships as discussed in the diode model of the previous section.

EG, the energy gap, depends on the type of semiconductor material, and its value is given by

1.11 eV for silicon

0.67 eV for germanium

0.69 eV for Schottky-barrier

CJC, the BC depletion capacitance, is a function of the voltage applied across the base-collector junction. *CJC* is defined as the capacitance value when the voltage across the junction is zero. This parameter may not appear explicitly on data sheets. In such a case, you can estimate it from the equation

$$CJC = Cobo * \left(1 - \frac{-VBC}{VJC} \right)^{MJC}$$

where

VBC is the reverse voltage applied across the base-collector junction.

VJC is the built-in potential of the BC junction.

MJC is the BC grading coefficient.

Cobo is the capacitance measured at *VBC*.

Cobo and *VBC* are given on most data sheets.

A similar relationship exists for the parameter **CJE,** which is the BE zero-bias depletion capacitance. If you cannot find it on a data sheet, estimate it from

$$CJE = Cibo* \left(1 - \frac{-VBE}{VJE} \right)^{MJE}$$

where

VBE is the reverse voltage applied across the base-emitter junction.

VJE is the built-in potential of the BE junction.

MJE is the BE grading coefficient.

Cibo is the capacitance measured at *VBE*.

RB, the zero-bias base resistance, is an important parameter that affects the transient response and the small signal analysis. It is also one of the most difficult parameters to measure accurately because of the error introduced by the emitter resistance. *RB* depends strongly on the operating point of the transistor. In the Ebers-Moll model, *RB* is assumed to be constant.

RC, the collector resistance, causes the decrease in the slope of the *I-V* curves in the saturation region for low values of *VCE*. It is assumed to be constant.

RE, the emitter resistance, is the resistance between the emitter region and the emitter terminal, and it is assumed to be constant. A typical value is about 1 ohm.

VAF, the forward Early voltage, models the effect of basewidth modulation due to the variations found in the BC space charge region. If it does not appear on a data sheet, you can estimate it from the I-V characteristics of the transistor using

$$VAF = \frac{IC}{hoe} - VCE$$

where *hoe* is the output admittance of the transistor (usually found on data sheets).

VAR, the reverse Early voltage, is similar to **VAF**, but with the collector and emitter leads interchanged. This parameter is important only when the transistor will be operated in the reverse-active region.

TF, the forward transit time, is used to model the excess charge stored in the transistor when the BE junction is forward biased. It is used to calculate the emitter diffusion capacitance. It can be estimated from the unity gain bandwidth, **FT** (the frequency at which the common emitter, no load, small signal gain is unit). The following equation is used in the estimate:

$$TF = \frac{1}{2 * \pi * FT}$$

TR, the reverse transit time, is the total reverse transit time used to model the excess charge stored in the transistor when the BC junction is forward biased and with $VBE = 0$. This is needed to compute the collector diffusion capacitance. It is estimated from the equation,

$$TR = Temp1/Temp4$$

where

$$Temp1 = TS/BR$$
$$Temp2 = IBF + IBR$$
$$Temp3 = (ICF/BF) + IBR$$

$Temp4 = LN(Temp2/Temp3)$

ICF is the forward collector current.

IBF is the forward base current.

IBR is the reverse base current.

BR is the reverse beta.

BF is the forward beta.

TR is typically larger than *TF* by one to three orders of magnitude.

MJC, the base-collector grading coefficient, is similar to the parameter *M* of the diode model. You need two sets of data points. These values can be obtained directly from the graph of *Cobo* vs. *VCB* found in most data sheets.

$$MJC = \frac{LN\left(\dfrac{Cj1}{Cj2}\right)}{LN\left(\dfrac{Vr2}{Vr1}\right)}$$

Cj1 is the capacitance measured at *Vr1*, and *Cj2* is that measured at *Vr2*.

VJC, the base-collector built-in potential, is the built-in potential of the base-collector junction. It can be obtained in a similar way as *VJ* in the diode model.

MJE and **VJE** follow similar relationships to *MJC* and *VJC*, only the voltage used in the equations is *VEB* instead of *VCB*.

MJS, the collector-substrate grading coefficient, and **VJS,** the collector-substrate built-in potential, are found in the same manner as *MJC* and *VJC* for the diode model. **CJS,** the collector-substrate depletion capacitance, follows the equation

$$CJS = CJS(VCS) * \left(1 - \frac{-VCS}{VJS}\right)^{MJS}$$

where

VCS is the reverse voltage applied across the collector-substrate junction (a positive value).

CJS is the zero-bias depletion capacitance.

VJS is the built-in potential of the *CS* junction.

MJS is the *CS* grading coefficient.

CJS(VJS) is the capacitance measured at *VCS*.

NF, the forward emission coefficient, is similar to the parameter *N* of the diode model.

NR, the reverse emission coefficient, applies to the reverse-active mode of operation.

ISE, the base-emitter saturation current, can be estimated from the DC beta vs. *IC* curve. It is typically larger than *IS* and is used in the Gummel-Poon model.

ISC, the base-collector saturation current, normally does not have a great impact on the collector current in normal forward-bias conditions.

IKF, the corner for forward beta high current roll-off, can be estimated from the beta vs. *IC* curve. It is usually defined by one-half of the maximum beta. It is valid at high collector currents. Typical values range from 10 mA to 100 mA.

IKR, the corner for reverse beta high current roll-off, is similar to *IKF,* and it is important only when you operate the transistor in the reverse-active mode.

NE, the base-emitter emission coefficient, can be obtained from the slope of the beta vs. *IC* curve in the low collector region where beta is still increasing with collector current.

NC, the base-collector emission coefficient, is similar to *NE* except that you reverse the collector and emitter leads. This parameter can be significant if you operate the transistor in the reverse-active mode.

IRB is the current level at which the base resistance drops to half of *RBM*.

RBM is the minimum base resistance at high currents. The base resistance reduces at high currents because of current crowding.

VTF, the *VBC* dependence voltage, is associated with the forward transit time. It is estimated from the following equation:

$$VTF = \frac{Vcez - Vcey}{1.33*LN[(Fty/Ftx)*(Ftmax - Ftx)/(Ftmax - Fty)]}$$

The required data can be obtained from *FT* measurements. It should be verified with actual simulation.

XTF, the coefficient for bias dependence of *TF,* is computed at a low value of *VCE* such that *VBC*=0. *IC* is set to a high value so that *FT* approaches minimum. This value of *FT* is defined as *Ftmin.* *XTF* is then computed from

$$XTF = \frac{Ftmax}{Ftmin} - 1$$

ITF, the high current parameter for effect on *TF,* is related to the forward transit time. It is estimated by setting *VBC* near 0 volts. The collector current, *IC,* can be found from the peak of the *FT* curve. Calling this value *Ft', we then find ITF* from

$$ITF = \frac{IC}{\sqrt{XTF*(Ftmax/Ft'-1)}} - IC$$

PTF, the excess phase, is computed at a frequency of

$$f = \frac{1}{2\pi TF}$$

It can be found by measuring the phase angle at the unity-gain bandwidth of the transistor. It is the difference in phase between *FT* and 90°.

XCJC models the fraction of the base-collector capacitance that is connected to the internal base node. It is important

XCJC models the fraction of the base-collector capacitance that is connected to the internal base node. It is important at high frequencies. The capacitance, XCJC*CJC is the capacitance connected between the internal base lead and the collector. Alternatively, (1 − XCJC)*CJC is the capacitance connected between the external base lead to the collector. *CJC* is the total capacitance.

XTI, the saturation current temperature exponent, describes the manner in which saturation current changes with temperature. The saturation current at temperature T2 is given by

$$IS(T2) = IS*\left(\frac{T2}{Tnom}\right)^{XTI} *e^{\,[qEg(300)/kT2](1\,-\,Tnom/T1)}$$

KF and **AF** are used only in noise analysis, which is a subset of AC analysis. The noise is computed for a 1 Hz bandwidth. The model handles thermal, flicker, and shot noise. The thermal-noise current is generated by *RE, RB,* and *RC,* and the mean square of the noise current is given by 4kT divided by the resistance as follows:

$$I^2 = \frac{4kT}{R}$$

Flicker- and shot-noise current is related to the collector and base currents and depends on frequency. The mean square noise current is given by

$$I^2 = 2*q*I + KF*I^{AF}/Frequency$$

where *KF* is the flicker-noise coefficient, and *AF* is the flicker-noise exponent.

FC, the coefficient for forward-bias depletion capacitance, is common to all the junction capacitors in the *BJT* model. It controls the turning point of *CJC, CJE,* and *CJS* when the junctions are forward biased. When the junction voltage exceeds *FC*VJ*, the depletion capacitance is replaced by the equation given for *CJS*.

Operational Amplifiers

The model of the operational amplifier contains 12 parameters, as follows:

Name	Parameter	Units
RIN	Input resistance	ohms
AO	Open-loop gain	—
ROUT	Output resistance	ohms
VOFF	Offset voltage	volts
TC	Temperature coefficient for VOFF	V/°C
F1	First pole location	Hz
F2	Second pole location	Hz
SR	Slew rate	V/sec
IOFF	Input offset current	amps
IBIAS	Input bias current	amps
DI	Current doubling interval	°C
VMAX	Maximum output voltage	volts

RIN, the input resistance of the op-amp, can be found on the data sheets. A typical value is 2 megohms.

AO, the open-loop gain, is found on data sheets. A typical value is 200,000.

ROUT, the output resistance, can be found in most data sheets. Typical values are 50 to 75 ohms.

VOFF, the offset voltage, is due to a mismatch at the input stage. This parameter appears on the data sheets. Typical values are in the millivolt range.

TC, the temperature coefficient of the input offset voltage, is found on the data sheets.

F1, the first pole, is the dominant pole of the operational amplifier. In most applications, it is the only important pole. It can be obtained from the data sheets or computed from the open-loop gain and gain-bandwidth product using the equation,

$$F1 = FT/AO$$

where FT is the frequency at which the gain is unity.

$F2$, the second pole, is important for high frequency effects. It is usually in the MHz range, and for applications where it is not needed, you can set it to 1E10.

SR, the slew rate, is the maximum rate of change at the output. It is an important parameter that limits the frequency that can be applied at the input. The units for slew rate are V/sec, and a typical value is 0.5 V/usec.

$IBIAS$, the input bias current, is found directly on the data sheet. It is usually higher when the op-amp uses bipolar input and much lower for op-amps with JFET inputs. It is defined by the following equation:

$$Ibias = \frac{Ip + \epsilon}{2}$$

where

Ip is the current flowing into the noninverting lead.

In is the current flowing into the inverting lead.

$IOFF$, the input offset current, is due to mismatch of the input devices. It is found directly on the data sheet.

DI, the doubling interval, is the temperature change for which Is and $Ibias$ double. This parameter has no effect if you are running a simulation at room temperature or when $Ibias$ and Is are set to zero.

$VMAX$, the maximum voltage, is the maximum allowable output voltage of the op-amp.

References

Semiconductor Device Modeling with SPICE, Paolo Antognetti and Giuseppe Massobrio, McGraw-Hill, 1988.

Small-Signal Transistor Data, Motorola Inc.

Device Electronics for Integrated Circuits, 2d ed., Richard S. Muller and Theodore I. Kamins, Addison-Wesley, 1986.

3

Quick Reference Keyboard Commands

Keyboard Commands

F1	Displays the help screen for the current window.
F2	Starts analysis run (if in an analysis mode).
F3	Exits analysis routine (if in an analysis mode).
	Exits the main program (if in the schematic editor mode).
F4	Brings the bottom window to the foreground.
F5	Toggles front window size between maximum and last size.
F6	Toggles between front window and Component window.
F7	Toggles between front window and Mode window.
F8	Scope (if in an analysis mode).

F9	Limits window (if in an analysis mode); toggles schematic scale (if in the schematic editor mode).
F10	Options window (if in an analysis mode).
F11	Parameter Stepping window (if in an analysis mode).
F12	Transient analysis Monitor window.
Alt-F1	Screen dump at large scale.
Alt-F2	Screen dump at small scale.
Alt-F3	Prints visible contents of front window at small scale.

Schematic Editor Keyboard Commands

Ins	Inserts the last selected object at cursor location.
A	Adds object at cursor location.
C	Switches element to Component.
Cursor keys	Move components before pressing ↵.
D	Defines box size or allows parameter change.
Del	Deletes object at cursor location.
I	Displays information on component at cursor location.
L	Switches element to Line.
M	Moves object to new cursor location.
R	Runs an analysis. The type of analysis is selected from the menu by typing the appropriate number (1 for transient, 2 for AC, 3 for DC, and 4 for Fourier).
S	Steps contents of box vertically or horizontally or both.

T	Switches element to Text.
Z	Deletes object at cursor location.
Space bar	Rotates component during Add and Move commands.
↵	Places component at location of cursor; marks end of line if line has been selected.
Esc	Erases component immediately after it has been added.

Index

Output voltage, 88–90
Output waveform, finding, 92–94

#Parameter statement, 50
Passive components, 34–37
   #Define statement and, 36
   #Model statement and, 37
Phase angle, plotting, 109
Phase range, AC analysis and, 106
Phase shift curve, 25, 26
Pivot solver, 64
PJFET transistors, 39
PMOS transistors, 38
PNP transistors, 38
Polynomial sources, 48, 61
Print interval, transient analysis and, 73
Printer
   problems with, 86, 115, 139
   setting up program for, 65
Printing
   of circuits, 63
   of front window, 63
   of graphics images, 63
   of netlist showing contents of, 63
   in numeric form, 73
   of simulation results, 86, 115, 139
Proportionality constant, dependent sources
   and, 48
Pulse sources, 41, 59–60

Quitting, exit verification before, 63

Real numbers, 35
Real range, AC analysis and, 105
Relational operators, v(t) sources and, 44
Relative error, transient analysis and, 73
Relaxation factor, 64
Resistance
   base, minimum at high currents, 176
   collector, 172

   emitter, 172
   off state value of, 64
   output, 178
   on state value of, 64
   zero-bias base, 172
Resizing, of windows, 55
Retrieval, of simulations, 76, 109, 135
Reverse beta, 171
Reverse early voltage, 173
Reverse emission coefficient, 175
Reverse transit time, 173–174
Ruler marks, 77
   AC analysis and, 110
   DC analysis and, 135
Run options
   AC analysis and, 109
   DC analysis and, 135
   transient analysis and, 75, 76

Saturation current, 171
   base-collector, 175
   base-emitter, 175
   temperature exponent and, 176
Saving
   of circuits, 21, 62
   of device library, to disk file, 62
   of simulation results, 86, 109, 114–115, 135, 139
   of waveforms, 146–147
Schematic editor, 31–67
   Add selection and, 33–53
   Box mode and, 53
   Comp mode and, 33–53
   Def selection and, 54
   Editors menu and, 56–61
   File menu and, 61–62
   Info selection and, 55
   keyboard commands for, 182–183
   Line mode and, 53
   Move selection and, 54
   μ menu and, 66

Addison-Wesley License Agreement